Agents and Goals in Evolution

Samir Okasha offers a philosophical perspective on evolutionary biology in *Agents and Goals in Evolution*. His focus is on 'agential thinking', which is a mode of thought commonly employed in evolutionary biology. The paradigm case of agential thinking involves treating an evolved organism as if it were an agent pursuing a goal, such as survival or reproduction, and treating its phenotypic traits as strategies for achieving that goal, or furthering its biological interests.

Agential thinking involves deliberately transposing a set of concepts—goals, interests, strategies—from rational human agents to the biological world more generally. Okasha's enquiry begins by asking whether this is justified. Is agential thinking mere anthropomorphism, or does it play a genuine intellectual role in the science? This central question leads Okasha to a series of further questions. How do we identify the 'goal' that evolved organisms will behave as if they are trying to achieve? Can agential thinking ever be applied to groups or genes, rather than to individual organisms? And how does agential thinking relate to the controversies over fitness-maximization in evolutionary biology?

In the final third of the book, Okasha examines the relation between the adaptive and the rational. If organisms can validly be treated as agent-like, for the purposes of evolutionary analysis, should we expect that their evolved behaviour will correspond to the behaviour of rational agents as codified in the theory of rational choice? If so, does this mean that the fitness-maximizing paradigm of the evolutionary biologist can be mapped directly to the utility-maximizing paradigm of the rational choice theorist? Okasha explores these questions using an inter-disciplinary methodology that draws on philosophy of science, evolutionary biology, and economics.

Samir Okasha is Professor of Philosophy of Science and Head of the Philosophy Department at the University of Bristol, where he has worked since 2003. He previously held positions at the University of York, LSE, and the National University of Mexico. Okasha is the author of numerous articles on topics in philosophy of science, philosophy of biology, evolutionary theory, and epistemology. His book *Evolution and the Levels of Selection* (OUP 2006) was awarded the Lakatos Prize for an outstanding contribution to the philosophy of science. He is currently President of the European Philosophy of Science Association.

Praise for *Agents and Goals in Evolution*

'*Agents and Goals in Evolution* is essential reading for philosophers and biologists interested in subjects where reference to agency occurs, including fitness optimization, kin selection, and social evolution. It also touches on the relation between rationality and evolution, which could make it of interest to scholars working outside evolutionary biology but seeking to understand the appeal to evolution in different scientific fields.'

Adrian Stencel, *History and Philosophy of the Life Sciences*

'His book is thought-provoking, and it provides an excellent entry point into an interesting multidisciplinary literature. I will certainly make use of it in the future as a reference work.'

Andy Gardner, *Metascience*

'Okasha's overall discussion is admirably clear, focused, and integrative, despite ranging over literatures from evolutionary biology, game theory, rational choice theory, and the philosophy of economics. He brings much order to what can be a confusing set of issues and debates . . . those looking for a clear overview of agential thinking in the evolutionary sciences will learn much from the book, and it will be important reading for philosophers of the biological and social sciences more generally.'

Robert A Wilson, *Notre Dame Philosophical Reviews*

'Samir Okasha's 2018 book might well become the consensus classic text for biologists to fall back on when they find themselves unable to resist both function talk and agent talk in the course of their inquiries and explanations. It covers the ground with admirable clarity, caution and scholarship, delving in detail into the formal work by Hamilton, Maynard Smith, Grafen, Trivers and others, while also considering a wealth of theoretical and empirical research in behavioral ecology, cognitive ethology, economics and psychology.'

Daniel C. Dennett, *Metascience*

'I have barely scratched the surface here of the many subtle, rich and illuminating points made in this book. Anyone with a serious interest in the foundations of evolutionary theory and the nature of evolutionary explanation will get a lot out of it, whatever their disciplinary background.'

Jonathan Birch, *Mind*

'remarkably well argued and deep for a book that covers so much ground. Okasha clarifies and organizes many formerly disparate ways of using agential thinking in biology, discussing grand ideas with extraordinary clarity and subtlety.'

Hannah Rubin, *Metascience*

Agents and Goals in Evolution

Samir Okasha

OXFORD
UNIVERSITY PRESS

OXFORD
UNIVERSITY PRESS

Great Clarendon Street, Oxford, OX2 6DP,
United Kingdom

Oxford University Press is a department of the University of Oxford.
It furthers the University's objective of excellence in research, scholarship,
and education by publishing worldwide. Oxford is a registered trade mark of
Oxford University Press in the UK and in certain other countries

First published 2018
First published in paperback 2020

Published in the United States of America by Oxford University Press
198 Madison Avenue, New York, NY 10016, United States of America

British Library Cataloguing in Publication Data
Data available

Library of Congress Cataloging in Publication Data
Data available

ISBN 978–0–19–881508–2 (Hbk.)
ISBN 978–0–19–289443–4 (Pbk.)

For Havi Carel

Contents

Detailed Contents

Preface and Acknowledgements

This is a book about evolutionary biology, written from a philosophical perspective. Its main concern is to analyse a mode of thinking in biology that is quite common, and philosophically interesting. I call it 'agential thinking', following Peter Godfrey-Smith. It assumes a variety of forms; but in its paradigm case, agential thinking involves treating an evolved organism as if it were an agent pursuing a goal, such as survival or reproduction, and treating its phenotypic traits, including its behaviour, as strategies for achieving that goal, or furthering its biological interests. This way of thinking might be thought uncontroversial, or at least no more controversial than the basic Darwinian assumption that organisms' evolved traits are often adaptive. But there is more to it than this. For agential thinking involves deliberately transposing a set of concepts—goals, interests, strategies—whose original application is to rational human agents, to the biological world more generally. What could be the justification for doing this? Is it mere anthropomorphism, or does it play a genuine intellectual role in the science?

This question is the starting point of my enquiry, but it leads to a series of further questions. How do we identify the 'goal' which evolved organisms will behave as if they are trying to achieve? Can agential thinking ever be applied to groups, rather than to individual organisms? How does agential thinking relate to the controversies over fitness-maximization in evolutionary biology? A further set of questions concerns the relation between the adaptive and the rational. If organisms can validly be treated as agent-like, for the purposes of evolutionary analysis, should we expect that their evolved behaviour will correspond to the behaviour of rational agents as codified in the theory of rational choice? If so, does this mean that the fitness-maximizing paradigm of the evolutionary biologist can be mapped directly to the utility-maximizing paradigm of the rational choice theorist? These are a sample of the questions addressed in the book.

The book's orientation is philosophical, but it adopts an interdisciplinary approach. It is written in the conviction that philosophy of science is at its most productive when done with a close eye on the science itself. As such, the book engages extensively with the evolutionary biology literature, and to a lesser extent with that of economics and rational choice. The book is aimed at philosophers of the biological sciences, evolutionists with a taste for conceptual issues, and interested parties from other disciplines. It assumes a basic familiarity with Darwinian evolution, but no specialist scientific or philosophical knowledge, and many concepts are explained from scratch. Parts of the book are somewhat technical—but no more so, I hope, than is necessary to tackle the issues properly.

The book is the result of many years' work. It would be impossible to acknowledge all of the people who have influenced my thinking, but particular mention must go to Ken Binmore, Elliott Sober, Kim Sterelny, Peter Godfrey-Smith, Daniel Dennett, Brian Skyrms, Alan Grafen, Andy Gardner, Arthur Robson, Cédric Paternotte, David Queller, Jonathan Grose, Tim Lewens, Jonathan Birch, Anthony Edwards, John McNamara, and James Ladyman. A number of colleagues provided valuable feedback on individual chapters: Alex Rosenberg, Alexander Bird, Richard Pettigrew, Tudor Baetu, Tim Lewens, Johannes Martens, and Bengt Autzen. I am particularly grateful to Ken Binmore, Nicholas Shea, and Patricia Rich, who provided detailed written feedback on multiple chapters, and to Jonathan Birch, one of two readers for OUP, who did likewise.

Early parts of the work were funded by a research grant from the Arts and Humanities Research Council, between 2008 and 2011. The majority of the research, and the actual writing of the book, was funded by a European Research Council Advanced Grant, agreement no. 295449, between 2013 and 2017. I am grateful to my employer, the University of Bristol, for allowing me the time necessary to bring the project to completion. I am grateful to Oxford University Press, John Wiley, and Springer for permission to re-use material published in *British Journal for the Philosophy of Science* vol. 59, *Journal of Evolutionary Biology* vol. 29, and *Biology and Philosophy* vol. 29, respectively. Finally, I am grateful to Havi, Solomon, and Joel for their endurance while I was preoccupied with this project, and to my parents who encouraged me to study philosophy in the first place.

List of Figures, Tables, and Boxes

Figures

Tables

Boxes

Introduction

There is a familiar story about the place of teleology in biology that goes as follows. Since Aristotle, biologists have used a teleological idiom to describe living organisms, but the justification for doing so only became apparent with Darwin. Though the process of evolution by natural selection is mechanical and lacks foresight, Darwinism nonetheless licenses talk of function and purpose in nature. In statements such as 'the polar bear's white coat is for camouflage' and 'the cactus has spines in order to deter herbivores', the teleological terms ('for', 'in order to') are really a way of talking about adaptive significance. Natural selection led polar bears to evolve white coats and cacti to grow spines because these traits helped to camouflage bears and protect cacti, so were adaptive. Thus Darwinism supplies a naturalistic basis for at least some of the teleological idioms that biologists had long used.

A related idea is that Darwinism placed teleological explanations on a respectable footing by showing them to be really causal. Explanations such as 'plants grow tall to gain more sunlight' appear to explain a feature by its effects rather than its causes; taken at face value, this involves either backwards causation or the attribution of conscious intent to plants, both of which are problematic. However, in the light of Darwin, we know that such explanations can be translated into purely causal terms. In the past, plants that grew tall obtained more sunlight than ones that didn't, so left more descendants, and thus the trait proliferated. That is, natural selection generates a feedback process in which a trait's effect causally influences its subsequent fate, thus showing the apparently teleological explanation to be causal in disguise.

The idea that Darwinism naturalizes teleology by identifying a trait's function with its adaptive significance has considerable merit. It makes sense of much biological usage, and provides a principled basis on which to determine when talk of function, design, and purpose is legitimate. (Though it is an open question whether all biological uses of the term 'function' should be understood this way.) Moreover, it helps explain, in a naturalistically acceptable way, why we apply a purposive idiom to living organisms and their traits, but not to mountains or rivers. Finally, the empirical commitments of the idea are fairly modest: that organisms exhibit traits which contribute positively to their Darwinian fitness, so have functions, is a commonplace of evolutionary biology.

My concern in this book is with a mode of thought in evolutionary biology that is related to the function-talk that Darwinism naturalizes, but is distinct from it.

It involves appeal to the notion of *agents* with *interests*, *goals*, and *strategies* in evolutionary analysis. I call this 'agential thinking', following Godfrey-Smith (2009). In the most common form of agential thinking, the agents are individual organisms, their goal is to survive and reproduce, and their evolved traits are strategies for achieving this goal. Behavioural ecologists often think about animal behaviour in these terms. In other cases, the entities that are treated as agents are genes or groups, rather than individuals. A still different form of agential thinking involves treating the process of natural selection itself, personified as 'mother nature', as agent-like, choosing between alternative phenotypes in accordance with a goal, such as improving a population or maximizing its fitness.

Agential thinking is a form of adaptationist reasoning—that is, of trying to understand evolved traits in terms of their contribution to fitness. As such, it is related to the sort of function-talk that can be straightforwardly naturalized. But it goes further, for it involves transposing a set of concepts—interests, goals, and strategies—whose original application is to the deliberate behaviour of human agents, to the biological world at large. When applied carefully, this can yield insight. For example, suppose we want to explain why a male rat tries to kill the pups produced by a female in their group, and why the female tries to stop him. By thinking of the male as an agent whose goal is to mate with the female, and who has devised a strategy for bringing her back to estrus, we can make sense of his infanticidal actions. Similarly, by treating the female rat as an agent with her own goal, we see immediately that male and female have different interests, thus explaining the conflict between them.

One symptom of agential thinking is the use of intentional language, such as 'knows' and 'wants', in evolutionary biology. Such language has its primary application in human psychology, but is often used in a biological context too, in an extended or metaphorical sense. Intentional language is surprisingly apt for describing and explaining adaptive behaviour, even of organisms with limited cognitive abilities, as Dennett (1987) observes. For example, consider a worker honey bee who eats the eggs laid by a fellow worker. This behaviour is adaptive, as it ensures that the offspring of the queen will predominate in the colony, which furthers the first worker's indirect evolutionary interests, given their close genetic relationship to the queen. Though the proximate cause of the worker's behaviour is chemical not psychological, it is extremely natural to explain the behaviour by saying that the worker knows that the eggs were laid by a fellow worker, and prefers that the queen's offspring are reared instead.

Though the use of agential and intentional idioms in evolutionary biology is both natural and familiar, from one perspective it is still quite puzzling. After all, it is not generally a useful strategy in science to treat the objects of one's study as if they had certain attributes which in fact they lack. Biologists do not find it useful to treat invertebrates as if they had backbones, after all. Why then would it be useful to treat evolved organisms as if they were agents pursuing goals, and to make them the subject of intentional attributions, when in fact they lack these attributes? What if anything is gained by thinking and talking this way, and how exactly does it relate to other ways of pursuing the adaptationist programme in evolutionary biology? Much of the first two chapters of the book is an attempt to answer these questions, though they recur throughout.

Agential thinking is intimately linked with the idea of *fitness-maximization* in biology. This idea has two variants, both controversial. The first is that evolved organisms will exhibit traits that are adaptive, hence maximize their fitness relative to some set of alternative traits. (How exactly 'fitness' should be defined is a major issue.) The second is that the process of natural selection itself involves maximization, in the sense of continually changing a population's composition so as to achieve higher fitness. These two claims are related but distinct. The former concerns adaptation (the product), while the latter concerns selection (the process). Both tie in with agential thinking, but of different sorts. The former relates to the paradigmatic sort of agential thinking in which the agent with the goal is an individual organism. The latter relates to the 'mother nature' sort of agential thinking, in which the agent with the goal is the evolutionary process itself. Thus to assess the validity of agential thinking, of either sort, we need to examine the status of fitness-maximization in biology. This is taken up in the middle chapters of the book.

Agential thinking is not just about the use of words, but also about models and explanatory strategies. Since the 1960s, concepts and models from rational choice theory have been used by biologists, with modifications, to help understand evolved behaviour. For example, Bayesian decision theory has been used to study animals' choices in the face of environmental uncertainty, such as the foraging strategies of birds (Valone 2006); signal detection theory has been used to explain aspects of animal communication (Wiley 2013); game theory was imported into biology from economics to shed light on animal conflicts, a development that gave rise to the field of evolutionary game theory (Maynard Smith 1982); and bargaining theory has been used to study the division of reproduction within animal societies (Cant and Johnstone 2009). In each of these cases, models originally designed to apply to rational human agents have been re-purposed for evolutionary analysis.

This transfer of ideas may seem surprising, given that most non-human organisms have limited powers of rational deliberation. What explains it? Part of the answer is that the concept of *utility* in rational choice and economics plays a somewhat similar role to the concept of *fitness* in evolutionary biology, as has often been noticed. Just as rational choice theorists assume that agents will behave in a way that maximizes their utility, so evolutionists often assume, and in some cases can show, that organisms will behave in a way that maximizes their Darwinian fitness (roughly, expected number of offspring), or some proxy for it. This conceptual link between utility-maximization and fitness-maximization was emphasized by Maynard Smith (1982). More recently, it features in the work of Grafen (2006, 2014a), who offers an explicit defence of the idea that an evolved organism can be modelled as an agent trying to maximize a utility function, where utility is suitably defined in terms of reproductive fitness.

Treating an organism as akin to a rational agent pursuing a goal is often heuristically valuable, and quite common in biological practice. Clearly, it is related to the function-talk that can be straightforwardly naturalized; for it is because organisms' behaviour is evolved, hence adaptive, that it can often be usefully assimilated to rational behaviour. But again, it goes beyond this point. For the uncontroversial idea that evolved behaviour is purposive in the sense of having a Darwinian function does not show that it corresponds, in any precise way, to the purposive behaviour of rational agents that is the subject matter of rational choice theory. Certainly there is a conceptual parallel here, and certainly the importing of rationality-inspired models into biology has borne intellectual fruit, but can a more explicit connection be forged between the two senses of purpose—that is, between the adaptive and the rational?

To make this more concrete, consider the problem of decision-making under uncertainty. Standard rational choice theory says that maximization of expected utility defines rational behaviour in this setting. It is tempting to suppose, as many authors have done, that by equating utility with Darwinian fitness, the same theory will define adaptive behaviour under uncertainty. But is this really true? Or consider social behaviour, that is, actions that affect others. In social contexts, an organism's personal reproductive success (fitness) is not the sole determinant of whether its genes will spread; relatives' reproduction matters too. So if utility is equated with personal fitness, then utility-maximizing behaviour will not be adaptive. A natural suggestion, often found in the literature, is that if utility is instead defined as 'inclusive fitness' in the sense of Hamilton (1964), then the link between adaptiveness and rationality can be restored. But is this true, and can a similar move be applied in other contexts? The answer is not obvious. So the idea that evolved organisms can be treated as akin to rational agents needs close scrutiny; this is taken up in the final four chapters.

This book is an extended reflection on agential thinking in biology, focusing on its rationale, its presuppositions, and the limits of its validity. Two overarching philosophical themes run through the book. The first is why agential thinking is so widespread. Is it a reflection of the objective biological facts, or of the human predilection to anthropomorphize? To see this contrast, consider the teleological attributions that succumb easily to Darwinian naturalization, such as 'the polar bear's coat is for camouflage'. There is a strong case for regarding these attributions as reflecting objective facts, or picking out a natural kind. Organisms really do exhibit adaptations, so their traits really are 'for' something. By contrast, consider Dawkins' concept of 'selfish gene' or Haig's 'strategic gene' (Dawkins 1976, Haig 2012). Both authors make a good case for the utility of these idioms in evolutionary genetics, and are clear about what they mean. Even so, the popularity of the idioms, and the mode of reasoning about evolution that accompanies them, is arguably a reflection of human psychology, at least in part, rather than the objective facts themselves.

The second philosophical theme concerns the relation between the use of intentional and rational attributions in evolutionary biology, and the actual evolution of intentionality and rationality. The point here is that the evolutionary process eventually gave rise to organisms, including humans, who have explicit beliefs, desires, and goals that are mentally represented, of whom intentional psychology is approximately true, and whose purposive behaviour can be codified, at least roughly, by the principles of rational choice theory. The cognitive machinery that underpins these capacities presumably evolved by natural selection. If so, this prompts a question. How should we make sense of the joint facts that intentional and rational idioms are used to theorize about evolution, and that there is an evolutionary story to be told about how the cognitive capacities needed for intentional and rational behaviour arose? How do these facts relate to each other?

Outline of the book

The book is organized into three parts, linked by a number of connecting threads. Part I explores the role of agential thinking in evolutionary biology, and its possible rationale. The notion of agency itself is explored, and a distinction is drawn between two types of agential thinking, both found in biological practice. In type 1, the agent with the goal is an evolved entity, typically an individual organism; in type 2, the agent is 'mother nature', a personification of natural selection. Agential thinking (type 2) is found to be misleading, applicable only to the simplest sort of natural selection. By contrast, type 1 is a legitimate expression of adaptationism, but it relies on a crucial presupposition. It presupposes that the entity that is treated as an agent exhibits a 'unity-of-purpose', in the sense that its evolved traits contribute to a *single* overall goal. Where this unity fails to obtain, as for example if there is substantial within-organism conflict, then agential thinking loses its grip: it becomes impossible to treat the organism as akin to an agent pursuing a goal. This is a biological analogue of the psychological unity-of-purpose that is presupposed when we attribute intentional states to humans. It explains the significance of agential thinking (type 1) but also its limitations. And it explains why it can only rarely be applied to groups rather than individual organisms.

Part II changes tack, turning to the long-standing debates over fitness-maximization in evolutionary biology. The idea that evolution by natural selection is in some sense an optimizing process, tending to maximize fitness, has a controversial status in biology. Some authors treat this as obviously true, others as demonstrably untrue except in special cases. The roots of this disagreement are explored. A number of attempts to make precise the idea that selection has an optimizing tendency are examined; all of them are related, with varying degrees of directness, to agential thinking. They include Wright's idea that selection will push a population up a slope in an adaptive landscape (Wright 1932); Fisher's 'fundamental theorem of natural selection' (Fisher 1930); Hamilton's idea that individuals' social behaviour will evolve to maximize their inclusive fitness (Hamilton 1964); Grafen's idea that individuals will evolve to be 'maximizing agents' (Grafen 2006); and the idea that frequency-dependent selection will lead to the evolution of traits that maximize fitness conditional on their being fixed in the population (Maynard Smith 1982). Each of these attempts is found to be only partly successful, which suggests that the link between natural selection and adaptation is weaker than is often assumed. This in turn helps to clarify the relation between the two types of agential thinking; it shows why type 2 is misleading, and shows that the justification for type 1 must ultimately be empirical, not theoretical.

Part III turns to the multi-faceted connection between evolution and rationality, in particular rational behaviour. This connection has two dimensions. The first is the idea that rationality, conceived of as an actual phenotypic attribute that some organisms including humans exhibit, may itself be an adaptation; the second is the idea that evolved organisms can usefully be treated as if they were rational, for the purposes of understanding their behaviour. The relationship between these two ideas, and the validity of each, is explored. Both depend on whether behaviour that is adaptive, or fitness-maximizing, coincides with behaviour that is rational, or utility-maximizing. This coincidence is often assumed in the literature, and sometimes treated as an a priori truth. But against this, there are also arguments that suggest that in particular contexts, the adaptive and the rational may part ways, in the sense that evolution may favour behaviours that *violate* the norms of rational choice theory. These 'parting-of-ways' arguments have a dual significance: they cast doubt on the assumption that rational behaviour is always biologically advantageous, and they suggest a limit on the use of agential thinking (type 1) to understand evolved behaviour. However, it turns out that in many cases, the coincidence between the adaptive and the rational can be restored by suitable choice of utility function.

The book employs a somewhat unusual methodology. Three different methods are used, interwoven with each other. The first is the traditional philosopher's technique of conceptual analysis—that is, trying to clarify the meaning of key concepts. The second is a method used widely in philosophy of science, of synthesizing a particular area of science then stepping back and asking, 'What does it all show?' It is suitable when the science in question is controversial, or has a hidden philosophical dimension. The third involves constructing simple formal analyses and models. This allows key ideas to be expressed more precisely, and the validity of specific arguments to be assessed. Though this combination of methods has its risks, my hope is that it allows the issues to be probed more deeply than by a single method.

Like most authors, I would ideally prefer that the book be read cover to cover, but realize that this is a lot to ask. There is extensive cross-referencing between chapters, but I have endeavoured to make each as self-standing as possible. Inevitably this entails a certain amount of recapitulation of earlier discussions, but this seemed a price worth paying. I recommend that all readers begin with chapter 1, which sets out the core problematic of the book. Thereafter, readers have a choice. Those primarily interested in evolutionary theory should continue on to chapter 2 and from there to Part II. However, it is also possible to skip from the end of chapter 1 straight to Part III; readers primarily interested in rationality may prefer this route.

PART I

Agency in Evolutionary Biology

1

Agential Thinking and its Rationale

1.1 Introduction

It is striking that evolutionary biology often uses the language of intentional psychology to describe the behaviour and activities of evolved organisms, their genes, and the process of natural selection that led to their evolution. Thus a cuckoo chick 'deceives' its host but will be evicted if the host 'discovers' that it is not its own; a worker ant 'wants' to raise the queen's offspring, not those of other workers; a swallow 'realizes' that winter is approaching so flies south; an imprinted gene 'knows' whether it was inherited paternally or maternally; and natural selection 'chooses' some phenotypes over others. This phenomenon has been aptly dubbed the 'cognitive metaphor' in biology by R. A. Wilson (2005, p. 75).

One might regard these intentional usages as unproblematic, simply a colourful gloss on biological facts that can be described in more neutral terms. The worker ant does not literally have wants but rather behaves as if it did, that is, it destroys eggs laid by other workers; the imprinted gene does not really know its origin but rather behaves as if it did, that is, it encodes a different phenotype depending on whether it is paternally or maternally inherited; and so on. Therefore uses of the intentional idiom in biology should be read in an 'as if' sense; they simply reflect the fact that organisms and genes are evolved entities and thus display or encode adaptive traits. It may be convenient to describe the activities of these entities in intentional-psychological terms, but in principle this could be avoided, and no particular theoretical significance attaches to it.

There is something right about this argument, but I do not think it is the whole story. For the intentional idiom in biology is a symptom of something deeper, namely a mode of thinking about Darwinian evolution that Godfrey-Smith (2009) has called 'agential'. This involves using notions such as *interests*, *goals*, and *strategies* in evolutionary analysis. One common form of agential thinking treats evolved entities as if they were agents consciously pursuing a goal, and had devised a strategy well-suited to achieving it. Behavioural ecologists studying the function of animal behaviour often think in these terms. For them, the agent is usually the individual organism and its goal might be to find a mate, protect its nest, survive the winter, or more generally to maximize its 'fitness' or some component thereof. In other cases, agential thinking is applied to whole groups, as for example in the idea that insect colonies display collective intelligence and make rational decisions (Edwards and Pratt 2009, Seeley 2010). In other cases the agents are taken to be genes or alleles, as in the 'selfish gene' concept of Dawkins (1982) or the 'strategic gene' concept of Haig (2012). Finally, Dennett (1987) applies agential thinking to the evolutionary process itself, treating 'mother nature', a personification of natural selection, as a rational agent who engineers solutions to the design problems faced by organisms.

I believe that agential thinking in biology, when used carefully, can be a powerful tool for understand adaptation. In life-history theory, for example, numerous aspects of an organism's life-cycle, such as the timing of reproduction or the length of its immature phase, can be understood by treating the organism as if it were an agent trying to maximize its expected number of offspring—or some other appropriate fitness measure—and had devised a strategy for achieving that goal. Or in social evolution theory, researchers have made sense of diverse social behaviours, particularly ones that involve altruism or self-sacrifice, by treating them as strategies used by an organism 'aiming' to maximize its inclusive fitness; indeed, this way of thinking is a major part of the motivation behind Hamilton's inclusive fitness concept (Hamilton 1964). Applied to genes, Dawkins (1982) makes a powerful case for the explanatory power of treating a gene as if it were a rational agent trying to devise ways to increase its representation in the gene-pool at the expense of its alleles. The phenomenon of intra-genomic conflict, in particular, makes good sense from this perspective.

There is an intimate link between agential thinking and the use of the intentional idiom ('knows', 'wants', 'tries') in evolutionary biology. For the language of intentional psychology applies in the first instance to human agents, who consciously have beliefs and desires, pursue goals, and choose actions appropriate to those goals. Thus a biologist who treats an evolved organism, a gene, or a group of organisms as akin to an agent with a goal will naturally be inclined to describe their activities in intentional-psychological terms.

The intentional idiom is one manifestation of agential thinking in biology, but it is not the only one. Another is the use of rational choice concepts in an evolutionary context, noted in the Introduction. The idea that what is adaptive, or fitness-maximizing, corresponds somehow to what is rational, or utility-maximizing, has long been mooted in the evolution of behaviour literature. Recently this idea has been developed by Alan Grafen, who argues that an evolved organism may be modelled as a rational agent seeking to maximize a utility function, or objective function. Grafen argues that this 'individual-as-maximizing-agent analogy' is a quite general way of thinking about adaptation, and is implicitly used by evolutionists in the field, but lacks a solid justification, which he hopes to provide (Grafen 2002, 2006, 2014a). In a similar vein, Elliott Sober (1998) describes what he calls the 'heuristic of personification' at work in biology; this heuristic tries to determine whether a trait will evolve by asking whether a rational agent seeking to maximize their fitness would choose that trait over alternatives. (Sober argues that the heuristic has limitations.) Grafen's and

Sober's ideas are analysed later in the book, in sections 4.2 and 7.2, respectively; for the moment, the point is just that they provide further evidence of the prevalence of agential thinking in evolutionary biology.

There are two possible attitudes that one might take towards agential thinking. The first sees it as mere anthropomorphism, an instance of the psychological bias which leads humans to see intention and purpose in places where they do not exist, and to favour teleological descriptions of the world over purely mechanical ones. This bias has been well-documented by psychologists. For example, Barrett (2004) has described a 'hypersensitive agent detection device' in humans, which leads us to make mistaken attributions of goal-directedness and intentional agency in the inanimate world; conceivably, a similar bias may be at work in biology. On this view, the explanation for the prevalence of agential thinking in biology lies in facts about human psychology, not in facts about the phenomena that biologists are trying to describe. A view of this sort is defended by Godfrey-Smith (2009), who argues that talk of agents and strategies, particularly as applied to genes, is a source of error in evolutionary biology; his arguments are discussed in section 2.2.2.

The second attitude sees agential thinking as a natural and justifiable way of describing or reasoning about the process of Darwinian evolution and/or its products. After all, many evolved organisms engage in activities that seem clearly purposive, such as foraging, searching for mates, and warning others of danger. Such behaviours are functionally similar to the actions of rational humans, even if their proximate cause is quite different, in that they are efficient means of achieving a goal (e.g. survival or reproduction). So treating the organisms in question as agent-like, and describing their activities in intentional terms, is well-motivated even if not literally true, in that it picks out a real phenomenon in nature. Similarly, one might try to justify (a different form of) agential thinking on grounds of an objective similarity between natural selection and rational choice: both are to do with selecting between alternatives in accordance with a goal, and thus involve a form of optimization. On this view, agential thinking in biology is not (mere) anthropomorphism, but has a genuine rationale and plays a real intellectual role.

My own attitude is intermediate between these two poles, as I think that agential thinking is not an undifferentiated whole. Partly this is because the notion of agency can itself be understood variously, as section 1.2 explains. Furthermore, there are two different ways in which the notion of agency, however understood, can be invoked in evolutionary analysis, depending on whether the focus is on selection (the process) or adaptation (the product), as section 1.3 explains. This leads to the key distinction between agential thinking of type 1 and type 2, both of which are found in biology; they are analysed in sections 1.4 and 1.5, respectively. Section 1.6 introduces *unity-of-purpose*, a key aspect of human agency, and argues that it has a biological analogue. This helps to rebut a possible objection to treating organisms as agents with goals, namely that it is merely a long-winded way of capturing the familiar point that evolved traits often have Darwinian functions. Section 1.7 examines a simple formalism, based on rational choice theory, for making agential thinking in biology precise. Section 1.8 looks briefly at Dennett's views. Section 1.9 concludes.

1.2 Concepts of Agency

To better understand agential thinking in biology, a good place to start is with the concept of agent itself. This concept is in fact understood differently in different fields; our first task is to describe briefly the variety of agent concepts and to consider how each applies to biology.[1]

The most minimal notion of agent, I suggest, is simply that of an entity that *does* something, or behaves. There is an intuitive distinction between an entity doing something and something happening to it, as Dretske (1988) has argued. To use one of his examples, consider the difference between a rat's moving its paw, which is something that the rat does, and a biologist moving the rat's paw, which is something that happens to the rat, not a behaviour of the rat. The difference, Dretske explains, lies in whether the proximate cause of the movement is 'internal' or 'external' to the rat. When the rat moves its paw, a series of neurological processes occurs in the rat's brain culminating in motor output. When the rat's paw is moved by a third party, the causal explanation is quite different, involving external forces acting on the rat. Though no doubt hard to make precise, Dretske's distinction is a real one and serves to define a minimal notion of agency.

Many though not all biological entities count as agents is this minimal sense. Cells divide, mitochondria make ATP, bacteria swim, plants climb, lions hunt, and insect colonies swarm. In these cases the proximate cause of what occurs is internal to the entity—though external factors may be necessary background conditions. To see this, contrast an insect colony's moving when it swarms with its moving when displaced by a hurricane. In the latter case, external factors wholly account for the movement; in the former, external factors, for example, ambient temperature suitable for swarming, are at most background conditions. Thus swarming is something that the colony does, not something that happens to it. An example of a biological entity that is not an agent is a species. Though we talk about a species 'going' extinct or 'producing' a daughter species, the active voice here is misleading. Extinction is something that happens to a species when all its members die, and speciation is something that happens to it when it is split into distinct sub-populations that diverge; these are not species' behaviours. Similarly, other taxonomic units such as clades are not agents either.

[1] Two usages of the term 'agent' should be mentioned only to set them aside, as they are unrelated to our concerns. The first is the use of 'biological agent' to mean a microbe suitable for use in biological warfare. The second is the use of 'selective agent' to mean ecological variable leading to differential survival, as for example when predation is said to be a selective agent for butterflies (e.g. Wade and Kalisz 1990). Here 'agent' simply means causal factor.

An interesting question is whether genes qualify as agents in the minimal sense. We say that genes replicate, and make proteins, but should such locutions be taken at face value? It is more accurate to say that genes are replicated by the cellular machinery—as opponents of gene-centric views of evolution have often stressed—so arguably replication is not something that a gene does. The same is true of protein synthesis: the causal process that leads from gene to RNA to protein is initiated and orchestrated by cellular activities, so involves factors external to the gene. However, if the event to be explained is not why a particular protein was synthesized in a cell at a particular time, but why the synthesized protein had the amino acid sequence it did, then the cause *is* internal to the gene: its DNA sequence. Perhaps this shows that instead of saying that genes 'make' proteins, we should really say that they determine protein primary structure—though whether the latter counts as something that a gene *does* is a moot point. I see no clear way of resolving this issue, but this should not obscure the fact that the minimal notion of agency has many clearcut cases on either side.

Philosophers have traditionally been concerned with a less catholic concept of agency, according to which agents are entities that *act*, rather than merely behave (Schlosser 2015). 'Act' here refers to intentional action, that is, doing something for reasons, or from intentions. How exactly this should be unpacked is controversial, but the standard view is that the action must have a particular aetiology—it must be suitably caused by the agent's psychological states, such as their beliefs and desires. Thus consider a typical human action, for example, opening the window in a stuffy room. The explanation of such an action will be something like: the agent wanted to let cool air into the room and believed that opening the window would achieve this. This notion of agency is thus intimately linked with the ability to engage in practical reasoning, that is, to work out how best to achieve one's ends, and with the concept of instrumental rationality—the type of rationality involved in choosing actions appropriate to one's ends, given one's beliefs.

Agents in this philosophical sense are required to have a specific psychological architecture, namely belief-like and desire-like states which give rise to actions. How widely this architecture is found among non-human organisms is controversial; but it is fairly clear that many biological entities, for example plants, genes, and microbes, will not qualify.[2] A bacterium that swims towards an oxygen gradient does not do so because it believes that the oxygen lies upstream and wants to get it; the correct explanation of its movement is non-psychological. Similarly, a gene which biases meiosis in its favour does not do so out of a desire to secure preferential access to the gametes. So the gene and the bacterium are not intentional agents and do not act, for they lack the relevant psychological states. I take it that this would be agreed on by all parties, even those sympathetic to Dennett's view that no sharp line distinguishes 'real' intentional systems from ones which it is convenient to treat as such (Dennett 1987).

[2] See Carruthers (2006) and Kornblith (2002) for a defence of the idea that belief-desire like architecture is widely found in the animal world.

Though most biological entities are not intentional agents, this concept of agency is still relevant to our concerns, for agential thinking in biology often involves treating entities as agent-like, not as literal agents. Clearly it could be useful, for predictive or explanatory purposes, to treat an entity that does not act for reasons as if it did, for example, if its behaviour is suitably similar to an intentional agent's. However, other disciplines have described alternative notions of agency, with weaker psychological requirements, relative to which many biological entities count as agents in more than an as-if sense. These notions are a half-way house between the minimal notion and full-blown intentional agency.

In A.I., an intelligent agent is defined as any entity that perceives or senses its environment and performs actions which alter the environment (Russell and Norvig 1995). Examples include simple control systems such as thermostats, software agents, and robots. The key attribute of agency in this sense is *flexibility*. An agent does not always do the same thing; rather, what it does depends on what it perceives (and sometimes on its inbuilt knowledge). The simplest intelligent agent is a 'reflex agent' whose action depends only on its current percept; it thus implements a set of stimulus-response conditionals. More sophisticated agents have an internal model of their environment which they update, so can learn from experience; they have a goal which they are trying to achieve; and in some cases they can engage in search and forward planning in order to achieve their goal. The behaviour of such agents is not merely flexible but also *goal-directed* and *autonomous*.

Many biological entities qualify as intelligent agents in this sense. Virtually all organisms, from microbes to animals, exhibit adaptive responses to environmental stimuli—think of a microbe swimming towards oxygen or a plant growing towards light. The same is true of sub-organismic entities such as cells and organelles, and collective entities such as honey-bee colonies and slime-molds. So flexibility is a common attribute of biological entities. Goal-directedness is also widespread. As Mayr (1988) noted, organismic activities such as foraging, seeking mates, and migrating are clearly goal-directed, not in the sense that the organism has a mental representation of the goal, but in that the activity is guided by an inbuilt genetic programme designed to achieve the goal; see section 1.5.1. Finally, many organismic subsystems, for example, the vertebrate immune system, also display flexibility and goal-directedness. So intelligent agents in this sense are abundant in biology.

A different notion of agency is found in the economics literature, in the rational actor model (e.g. Kreps 1988). In this field, a rational agent is defined as one whose decisions or choices maximize their utility, or expected utility in the case of decision under uncertainty. Utility-maximization is not intended as a psychological description but rather as a behavioural characterization: it means that agents behave *as if* they were trying to maximize a utility function. (In effect, this is a de-psychologized version of the notion of agency as intentional action.) In the simplest case of choosing between certain options, then as long as an agent's binary choices meet simple consistency conditions, such as transitivity, then the agent behaves as if they have a real-valued utility function on the options and always prefer the option with the highest utility. More complicated cases work in essentially the same way: the agent's choices or preferences are assumed to meet consistency conditions, which then imply the existence of a utility function which the agent behaves as if they want to maximize (section 1.7). Thus agency in this sense means rational pursuit of a goal, which boils down to consistency of choice.

The rational actor model was designed to describe human economic behaviour, but given its psychological neutrality it can apply to non-humans too. Any organism that can choose between alternatives, or make decisions, is potentially a rational agent. Exactly what 'choice' amounts to is not entirely obvious, but it is clear that many organisms with nervous systems, even simple ones, are capable of making within-lifetime choices, despite their lack of psychological sophistication. Thus butterflies choose what plant to oviposit on, birds choose whom to mate with, and primates choose whom to groom. Indeed, researchers have studied whether the choice behaviour of rats, pigeons, and even insects satisfies the axioms of the rational actor model, such as transitivity of choice; for the most part, it does.[3] Thus agency in this sense is found quite widely in biology.

To summarize, we have distinguished four notions of agency: the minimal notion of doing something; the philosopher's notion of agency as intentional action; the A.I. notion of agency as flexible/goal-directed activity; and the economist's notion of agency as rational choice. The minimal notion and the A.I. notion apply widely in biology, the philosopher's notion the least widely, while the economist's notion has intermediate generality. This refers to the *literal* application of each notion; however, the notions can also be applied metaphorically. We shall see that agential thinking in biology incorporates elements of all four notions, with an admixture of literal and metaphorical usages.

1.3 Two Types of Agential Thinking

Agential thinking in evolutionary biology comes in two types. The first type treats an actual evolved entity, paradigmatically an individual organism but possibly a gene or group, as akin to an agent with a goal. Thus for example a behavioural ecologist might treat an organism as 'trying' to survive and reproduce, and its phenotype as a strategy for achieving that goal, or for achieving intermediate goals such as mating or acquiring food. The second type treats 'mother nature', a personification of natural selection, as an agent who chooses between phenotypes in accordance with a goal, such as fitness-maximization; or (in another version) who tries to solve design problems. Here it is the evolutionary process itself that is treated as agent-like and made the subject of intentional attributions.

[3] See for example Kagel *et al.* (1995), Kalenscher and van Wingerden (2011), or McFarland (2016).

Both types of agential thinking involve a form of teleology, or goal-directedness, but in different ways. In type 1, the *telos* belongs to an evolved organism (in the paradigm case); the point of treating the organism as agent-like is to capture the fact that its evolved traits, including its behaviour, are adaptive, hence conduce towards the goal of survival and reproduction. In type 2, by contrast, the putative *telos* belongs to the evolutionary process itself ('mother nature'); the suggestion is that natural selection has an inherent tendency to move the population in a particular direction, so is goal-directed in that sense. Thus in the former case the teleological description applies to the *products* of evolution, while in the latter case it applies to the evolutionary *process* itself.

In effect, this means that the distinction between agential thinking of types 1 and 2 corresponds to the distinction between adaptation (the product) and natural selection (the process). On the usual view of the matter, there is a close link between selection and adaptation: the former explains the latter. So one might think that the two types of agential thinking stand or fall together. However, in the course of this book, I argue that this is not so. Agential thinking (type 1) is broadly defensible—it is a legitimate expression of adaptationist assumptions, and so is applicable quite widely. Moreover, it plays a real role in the science and can often lead to insight, though it must be used with care. But agential thinking (type 2) is more problematic—though it too can yield insight, it can equally easily mislead, and is an inappropriate metaphor for many forms of natural selection. In the next two sections I consider the two types of agential thinking in turn, in reverse order.

1.4 Mother Nature as an Agent

Natural selection is daily and hourly scrutinising, throughout the world, every variation, even the slightest; rejecting that which is bad, preserving and adding up all that is good; silently and insensibly working . . . at the improvement of each organic being.

C. Darwin, *The Origin of Species* (1859, p. 133)

Darwin himself was the first to employ agential thinking (type 2). In *The Origin of Species*, he frequently described natural selection as if it involved a conscious agent pursuing an agenda, as in the passage above and many similar ones. Darwin's choice of language was deliberate, and designed to emphasize the parallel between natural and artificial selection. Just as an animal breeder consciously chooses organisms with the desired attributes, so 'Nature' chooses organisms with fitness-enhancing attributes. Darwin admitted that he found it 'difficult to avoid personifying the word Nature', but argued that 'such metaphorical expressions' were harmless and 'almost necessary for brevity' (1859, p. 135). Darwin dismissed objections on this score as 'superficial', arguing that it was perfectly clear what his metaphors meant.

Darwin was perhaps too cavalier about this. Though the parallel with artificial selection aided his own route to the theory, the language of conscious agency made it harder for readers to grasp the crucial point that natural selection is a blind, mechanical process, lacking foresight, and does not unfold according to any plan. It is an interesting historical question whether Darwin's language hindered correct understanding of his theory. Be that as it may, today the point that natural selection is blind and mechanical is of course well-understood, and routinely impressed upon students of evolution.[4] We still speak of natural selection 'favouring' some variants over others, but such locutions surely cause no confusion today, even if they may have done in Darwin's time.[5]

This prompts the questions of whether agential thinking (type 2) plays any real role, explicit or implicit, in our modern understanding of evolution, and whether it should do. Perhaps surprisingly, I think that it *does* play a role, though not an entirely happy one. To tackle this issue, I want to explore three possible motivations for this type of agential thinking; the first two concern the nature of selection, the third the nature of Darwinian explanation.

1.4.1 Natural selection as rational choice?

The first motivation is that there is a non-trivial parallel between the choices of a rational agent and the 'choices' made by natural selection between alternative phenotypes or genotypes. Just as a rational agent, in the economic sense described above, prefers options that bring higher utility, so natural selection prefers alternatives that bring higher fitness. Thus both processes involve a form of optimization: choosing the 'best' member of a set at a given time. Whenever selection operates on a biological population, the set of alternative phenotypes can be ordered by their fitness in the current environment; if we wish, we can treat this as the preference order of a fictitious agent, namely mother nature. This allows us to rationalize selectively-driven changes in a population's composition in terms of mother nature's preferences, just as an agent's observed choices can be rationalized in terms of their preferences, in standard rational choice theory.

Why might this parallel be thought non-trivial? One answer is that it has played an actual role in science, for it partly underpins how game-theoretic ideas came to be introduced into biology (Maynard Smith and Price 1973, Maynard Smith 1982). In classical game theory the players are rational agents who aim to maximize their utility. The players choose between alternative strategies in real time, with consequent effects on their payoffs, that is, changes in utility. In biological game theory, in its simplest version, the players are organisms with hard-wired strategies. The organisms engage in pairwise social interactions with consequent effects on their payoffs, that is, changes in biological fitness; as a result, some strategies spread and others decline. Thus natural

[4] In a well-known textbook, D. Futuyma (2009) writes: 'natural selection is not an external force or agent, and certainly not a purposeful one' (p. 284).

[5] For a rare exception to this generalization, see Fodor and Piattelli-Palmarini (2010).

selection plays an agent-like role, choosing between strategies in accordance with expected payoff.

Maynard Smith (1982) observed that by replacing rational agency with natural selection, much of classical game theory could be given a biological re-interpretation. As he put it, 'the criterion of rationality is replaced by that of population dynamics and stability, and the criterion of self-interest by Darwinian fitness' (p. 2). The basic solution concept of classical game theory is the Nash equilibrium, which describes a situation (i.e. strategy profile) from which no rational player has a unilateral incentive to deviate. In its place, Maynard Smith devised the evolutionary stable strategy (ESS); this is a strategy which, if fixed in a population, cannot be invaded by other strategies. Though the former is an equilibrium in rational deliberation and the latter an equilibrium of an evolutionary process, there is a close mathematical relation between them,[6] and many theorists regard this as indicative of a deep conceptual link.[7]

This suggests that analogizing natural selection to agential choice cannot be dismissed as a triviality. On the other hand, more recent developments in evolutionary game theory partly undermine the rational choice/natural selection parallel. Maynard Smith's original approach did not explicitly study the evolutionary dynamics, but rather focused on characterizing the presumed stable endpoint of the evolutionary process. However, more recent work has shown that depending on model assumptions, the evolutionary dynamics may be cyclical, never settling down to an equilibrium at all; that even if an ESS exists, natural selection will not necessarily drive a population towards it; and that in certain cases, natural selection may drive a population to an unstable 'branching point', which is not an ESS at all (nor a Nash equilibrium), and at which (individual) payoff is actually *minimized*. These phenomena are discussed in section 4.3; for the moment, the point is that they were uncovered only by going beyond the focus on equilibrium that derived from rationality-inspired analysis.

Moreover, the complexities of evolutionary dynamics in game-theoretic scenarios suggest that personifying natural selection is not always particularly apt, and may mislead. Though it is always formally possible to interpret the ordering of a set of alternatives by their fitness, in a given environment, as a fictitious agent's preference ordering at a given time, we expect a real agent to have *stable* preferences. However, in game-theoretic scenarios the fitnesses of the different phenotypes (or strategies) is frequency-dependent, so the selective environment is always changing; thus the fictitious agent's preferences change too. It is as if mother nature continually chooses the phenotypic alternative she most prefers, only to find that her tastes have changed a moment later, so she needs to choose again; and this process can continue indefinitely.

[6] The ESS is a logically stronger concept: every ESS is a Nash equilibrium, though not vice-versa. But if a strategy is a strict Nash equilibrium, it is necessarily an ESS. See Nowak (2006) chapter 2 for a clear discussion of this.

[7] For example, Easley and Kleinberg (2010) pp. 209–10, and Hart (2005).

Viewed at a single point in time, mother nature's choices may resemble those of a rational agent; but viewed over time, she looks more like a fickle child.

The upshot is this. It is true that there is an abstract parallel between natural selection and rational choice—both involve choosing from a set of alternatives in accordance with maximal fitness/utility—which has played a role in science. But the parallel must be treated with care, and can easily mislead, for it licenses no simple prediction about the outcome of the evolutionary process except in cases where the selective environment remains constant over time.

This tallies with the conclusion Sober (1998) draws from his analysis of the 'heuristic of personification' in biology, though for different reasons. That heuristic, to recall, says that a given trait will evolve by natural selection if and only if a rational agent, trying to maximize their fitness, would choose that trait over the alternatives. Sober argues that the heuristic often works well, but breaks down in game-theoretic scenarios. His main example involves a Prisoner's Dilemma scenario (with additive payoffs); the rational agent chooses defection but natural selection favours altruism, owing to correlation in the population. However, in this example the personification heuristic would actually work perfectly well if the rational agent's goal was to maximize their *inclusive* rather than personal fitness; see sections 5.3 and 7.2. But in more complex game-theoretic scenarios, such as those involving cyclical dynamics and branching points, it is true that the evolutionary outcome cannot be predicted by asking what strategy a rational agent would choose, even if the agent's goal is suitably specified.

1.4.2 Natural selection as goal-directed?

The second possible motivation for agential thinking of the 'mother nature' variety is that natural selection mimics what a conscious agent, deliberately pursuing the goal of fitness-maximization, would do. So although natural selection is not in fact a goal-directed process, it nonetheless appears like one. This is suggested by Darwin's remark quoted above, that natural selection works 'silently and insensibly . . . at the improvement of each organic being' (1859, p. 133). Unlike the first motivation, which says that natural selection and rational choice are formally similar, the suggestion here is that the effects of natural selection are indistinguishable from what would be expected if a conscious agent, with an explicit goal, were orchestrating the process.

This line of argument is rarely given as an explicit justification for the 'mother nature' metaphor, but I think it is in the background of much evolutionary discussion; for the conception of natural selection on which it rests is widespread. In the twentieth century, the idea that natural selection tends inexorably towards 'improvement' found expression in Sewall Wright's adaptive landscape metaphor, which exerted a potent influence on subsequent evolutionary biology. Wright (1932) argued that natural selection would lead gene frequencies to change in such a way that mean population fitness was maximized; so evolution by natural selection could be depicted as movement up a mean fitness gradient in a landscape, or 'hill-climbing'. On this view, natural selection has an inherent directional tendency, so mimics a process which is goal-directed, or unfolding according to an agent's plan.

The appeal of the adaptive landscape notion is easily understood. Since selection involves 'fitter' variants prospering at the expense of the less fit, surely it *must* involve hill-climbing? However, this is in fact only true in simple cases, in which the selective environment, in the relevant sense, remains constant, and the complexities of inheritance are ignored. In general, natural selection need not lead mean population fitness to increase, even with frequency-independent fitnesses. This issue is examined in sections 3.2 and 3.3. For the moment, the key point is to distinguish the action of natural selection at a point in time from the evolutionary consequences at later times. At any point in time, selection does indeed favour fitter variants, but the effect on mean population fitness depends on the environment, which itself changes because of selection. So 'mother nature' is continually striving for a goal but not necessarily getting any nearer it.

The upshot, I think, is that the second motivation for personifying natural selection is not a good one, in that it would only be valid if Darwinian evolution were an optimizing process in a stronger sense than it actually is. (It is true that at any time, selection chooses between alternatives by the criterion of maximal fitness; but over time this process does not necessarily maximize any quantity.) This is not to say that selection never acts in a directional matter, nor drives populations up adaptive peaks, nor leads to adaptation, but only that there is no theoretical guarantee of this. Indeed, in some cases natural selection can have long-term consequences that are detrimental for a population, such as driving it to the brink of extinction, reducing population fitness, or causing its density to fall so low that genetic variation is lost. Such untoward outcomes, even if relatively rare, belie the idea that natural selection mimics the action of a conscious agent pursuing the goal of fitness-maximization.

Importantly, the idea that evolution by natural selection is a directional process, tending towards the goal of maximum fitness, must be sharply distinguished from the idea that well-adapted organisms, whose phenotypes have been shaped by selection, often behave as if their goal was to maximize fitness. The latter conception—in which the agent is the evolved organism, not 'mother nature'—can often be justified on empirical grounds; the former, on the other hand, is a theoretical claim about how natural selection works which is not generally true.

The infirmities of the hill-climbing view of evolution are fairly well-known, but its popularity endures. Thus there is a disconnect between what evolutionary theory teaches about how natural selection works and what many people, in biology and beyond, assume about it. (Rice (2004) describes the idea that natural selection leads to maximization of some quantity as 'one of the most widely held popular misconceptions about evolution' (p. 37).) The roots of this disconnect are complex; however, I suggest that implicit agential thinking of the 'mother nature' variety may be partly responsible. If one assumes that natural selection's cumulative effects will mimic those of a conscious agent trying to 'improve' a population, one will naturally be led to think of Darwinian evolution in hill-climbing terms. If this is right, then personifying natural selection may be a source of error, not insight. This issue is explored further in section 3.2.

1.4.3 Reading mother nature's mind?

The third possible motivation for agential thinking of the 'mother nature' type concerns the nature of Darwinian explanations, that is, ones which explain traits in terms of their adaptive significance. Daniel Dennett (1987, 1995) has argued persuasively that such explanations should be thought of as deciphering mother nature's reasons for designing organisms as she has. Dennett's guiding idea is that evolved traits, including behaviours, have what he calls 'free-floating rationales'. For example, the rationale for a swallow's seasonal migration is avoiding cold weather: this is why (in the ultimate sense) the swallow migrates. However, the swallow itself is blindly following its impulses, not acting for reasons, so the rationale in question is not the swallow's own; rather, it belongs to 'mother nature'. Dennett is thus led to the striking thesis that Darwinian explanation involves taking the intentional stance towards mother nature, that is, reading her mind.

This way of injecting agency into evolutionary biology may seem eccentric, and one might reasonably ask what its point is. In fact I think that Dennett is on to something important, but that it is better captured in a different way; see section 1.8. For now, note that Dennett's idea does not automatically fall prey to the objection above—that natural selection need not lead to improvements in a population. This objection points to (one) reason why adaptationist explanation has its limits, but that is compatible with Dennett's position. For the idea that such explanations involve 'reading mother nature's mind' says nothing in itself about how widely this mode of explanation can be applied.

To summarize, there are three possible motivations for agential thinking (type 2). The first is that natural selection and rational choice are formally similar processes; the second is that natural selection's effects mimic those of a conscious agent trying to achieve a goal; the third is that adaptationist explanation is usefully regarded as reading mother nature's mind. The first is valid in an attenuated sense though potentially misleading; the second is invalid if taken as a general claim about how selection works; and we have suggested that the third is under-motivated, though the argument for this has yet to be given.

1.5 Organisms as Agents

In agential thinking (type 1), it is an actual biological entity, not natural selection, that is treated as an agent with a goal and made the subject of intentional attributions ('wants', 'tries', 'prefers'). Here I assume that the entity is an individual organism. Other possibilities—genes and groups—are examined in chapter 2.

What is the point of treating an evolved organism as agent-like, in any of the aforementioned senses of 'agent'? I think we can distinguish three possible rationales, each of which is found in the literature. The first is that organisms are the locus of goal-directed activities. The second is that organisms exhibit behavioural flexibility. The third is that organisms have traits that are adaptations, so appear designed for a purpose. These rationales are not exclusive, and to some extent are complementary.

1.5.1 Goal-directedness

Many biologists have called attention to the goal-directedness of living organisms. Thus Mayr (1988) wrote: 'goal-directed behaviour . . . is extremely widespread in the organic world; for instance, most activity connected with migration, food-getting, courtship, ontogeny, and all phases of reproduction is characterized by such goal orientation' (p. 45). In a similar vein, Monod (1973) said that 'one of the most fundamental characteristics common to all living thing [is] that of being endowed with a project or purpose' (p. 9), while Waddington (1957) wrote that 'most of the activities of a living organism are of such a kind that they tend to produce a certain characteristic end result', which he referred to as 'directiveness' (p. 2).

The phenomenon described by these authors is unquestionably real. As characterized by Mayr, goal-directedness applies both to activities of the organism, for example, foraging and migration, and to processes that occur within an organism, for example, gastrulation and gametogenesis. In both cases the activity or process is initiated and guided by a 'genetic programme' which encodes the goal, Mayr argues. I think Mayr is right about this (despite some authors' qualms about the notion of genetic programme). These activities and processes are clearly genetically encoded (though in some cases involve learning too), and are goal-directed in a precise sense. They involve an orchestrated sequence of stages that can be observed repeatedly; they have a clear endpoint after which they cease; and the endpoint is reliably achieved in spite of perturbing factors, thanks to feedback mechanisms and other compensatory adjustments.

The goal-directedness of an organism's activities make it natural, indeed almost inevitable, to treat the organism as agent-like.[8] Consider a biologist observing the courtship ritual of a bird of paradise, or a sea turtle's homing, or an ant following a pheromone trail. The first question the biologist will ask is what the bird, turtle, or ant is *doing*; without an answer, they do not understand what they see. Answering this question requires identifying the relevant goals: the bird's goal is to attract a mate, the sea turtle's to return to its birthplace, and the ant's to find food. (We could equally say that the bird is 'trying' to attract a mate, etc.) Notice that in locutions of this sort, we describe the goal of the *activity* by saying what the *organism's* goal is in performing the activity. This subtle semantic shift is usually harmless; it shows how easily we slip into

[8] Goal-directedness is one of the reasons behind Denis Walsh's recent defence of the idea that organisms are 'agents of evolution'; see Walsh (2015).

treating the organism as agent. Note also that the semantic shift applies less naturally to internal goal-directed processes; we do not say that an organism 'tries' to gastrulate, or to produce gametes.

When our biologist explains the activity of the bird, turtle, or ant by citing its goal, they employ a mode of explanation that has a close parallel in human affairs. We frequently explain a person's action by saying what their goal or intended goal is, for example, as when we say that someone is tiptoeing to avoid disturbing the neighbours. In the human case, of course, the goal is typically mentally represented, and part of the agent's reason or motivation for performing the action; that is not so in the animal cases above. But despite this, there is a similarity between the two explanatory modes, which invites the deliberate assimilation of one to the other. We thus find it natural to describe an organism that performs a goal-directed activity as if it were an agent consciously pursuing a goal. This involves anthropomorphism but of a well-motivated sort, given that goal-directedness is a real phenomenon in nature.

This first rationale for treating organisms as agent-like is implicit in how biologists talk, both in the usually unnoticed shift from goal of activity to goal of organism, and in the vocabulary commonly used to describe an organism's activities. Terms such as 'fleeing', 'warning', and 'hunting', which are predicated of whole organisms, contain a clear imputation of goal-directedness, as has often been noted (Wright 1976, Nissen 1997). Indicative of this is that it makes sense to ask whether an organism performing one of these activities has been successful. I want to make two points about this rationale for agential thinking.

Firstly, much goal-directed behaviour, for example, insect phototaxis, does not involve cognition. Despite this, we often describe such behaviour by saying that the insect is 'trying' to reach the light. But it would be odd to describe the insect as having a belief about where the light lies; its behaviour is too limited for this locution to be useful, even metaphorically. So this involves what might be called the 'minimal' intentional stance: we explain the organism's activity by saying what it is trying to achieve, but without attributing to it beliefs or instrumental rationality. A rough parallel here is with the sort of intentional attribution we make to pre-verbal infants; we say that a crying baby is trying to attract its mother's attention, but not that it believes that by crying it will get its mother will come.

Secondly, the goal of an organismic activity, in Mayr's sense, is distinct from its adaptive significance. We can discover the goal of the turtle's swimming—returning to its birthplace—without knowing why turtles do this, in the ultimate (evolutionary) sense. Thus when a biologist explains a (token) turtle's behaviour by saying that it is trying to reach its birthplace, this is not to give an ultimate explanation. Rather, it is to say that the behaviour arises from a genetic program with a particular endpoint. This is a proximate explanation, or a gesture towards one; notwithstanding the fact that the genetic programme itself is evolved, and has an adaptive function. Therefore this notion of 'goal' is distinct from the notion at work in some evolutionary discussions, in which organisms are said to have the 'goal' of maximizing fitness. In principle an organism might exhibit goal-directed activity that does not advance its evolutionary goal; but empirically this is the exception not the rule.

1.5.2 Behavioural flexibility

The second rationale for treating organisms as agents, or agent-like, is behavioural flexibility. This rationale underpins the use of intentional language in cognitive ethology, and also the use of Bayesian models as tools for describing and theorizing about animal behaviour.

Most organisms, even quite simple ones, are able to sense environmental stimuli and adjust their behaviour in response. Much stimulus-response behaviour is purely instinctive and does not involve information-processing or learning. But there is a continuum from such simple behaviour to the more sophisticated behaviours made available by complex nervous systems, especially in vertebrates, which involve learning, memory, and inference. Such cognitive processes greatly increase the flexibility of an organism's behavioural repertoire, facilitating adaptive behaviour across a wide range of circumstances. Non-human organisms thus often appear to behave in quasi-rational ways, exhibiting sensitivity to environmental cues and adjusting their behaviour to suit the environment they are in, or think they are in.

Where an organism exhibits sufficient behavioural flexibility, it can be tempting to explain its behaviour from the intentional stance, that is, in belief-desire terms, in just the way we explain human actions. Cognitive ethologists often succumb to this temptation, and some explicitly defend the propriety of doing so.[9] Thus for example Ristau (1991) has studied piping plovers that feign injury when a predator approaches. She argues that the best explanation of the plover's behaviour is that it wants to lead the intruder away from its young; only this accounts for the precision and timing of its actions. Similarly, Clayton *et al.* (2006) study the food-caching behaviour of scrub jays. The jays not only store and retrieve food, but also use strategies to reduce the chance that their food is pilfered, such as delaying caching if other birds are watching, and choosing locations that are concealed from others' view. Clayton *et al.* insist that the jays' behaviour should be explained by attributing to them beliefs, desires, and memories, arguing that alternative non-intentional explanations fail.

Note that behavioural flexibility of the sort that invites belief/desire attribution is distinct from goal-directedness, and taxonomically more restricted. Organisms that exhibit such flexibility will typically perform goal-directed activity—the plover and jay behaviours above are clearly goal-directed—but the converse is not the case. As we have seen, many goal-directed activities are purely instinctive and not very rational-like; at most, they invite description using the minimal intentional stance ('trying') rather than the full one ('believes and desires').

[9] See Allen and Bekoff (1999) for a good discussion of this issue.

The view of the cognitive ethologists above—that the birds in question are real intentional agents with belief and desire-like internal states—is not universal. An alternative view is that this is only 'as if' intentionality, not the real thing, for example, because the birds lack language, or conscious awareness, or sufficient inferential ability. A third view says that there is no sharp distinction between genuine and 'as if' intentionality in the first place, so the question of whether the birds are really intentional agents, or merely usefully treated as such for predictive purposes, is not a good one (Dennett 1987). This is a much-debated issue to which we cannot do justice here. Instead I want to make two points about the second rationale for treating organisms as agents.

Firstly, this rationale is sometimes found in conjunction with a weaker notion of agent. It is quite common for researchers who model animal behaviour to treat organisms as Bayes-rational agents.[10] For example, a foraging bird may have some prior information ('beliefs') about the distribution of food across patches in its environment, acquired either genetically or from prior experience, or both. As the bird begins to forage in a given patch, its success rate provides evidence about the quality of the patch, which it integrates with its prior information using Bayesian updating to form an updated estimate of the patch's quality, which then informs its decision about whether to stay or leave. There is some evidence that animals really can implement such Bayesian calculations (Valone 2006). In this research tradition, behavioural flexibility is again the rationale for treating an organism as agent-like, but the relevant notion of agent is rational economic agent, rather than intentional agent.

Secondly, in so far as behavioural flexibility is the motivation for treating organisms as agent-like, this again concerns proximate not ultimate explanation. Much flexible behaviour is of course adaptive, and the cognitive processes that underpin it are themselves Darwinian adaptations, presumably. However, the fact that an organism exhibits sophisticated enough behaviour to warrant a belief-desire explanation, or to make such an explanation predictively useful, or to justify modelling it as a Bayesian agent, is a non-historical fact about it. In principle this motivation would survive even if creationism turned out to be true.

1.5.3 Adaptedness

The third rationale for treating organisms as agents is that they exhibit adaptations, that is, traits that confer a fitness advantage and have evolved for that reason. This fact motivates the search for adaptationist explanations in biology, which try to identify the adaptive significance (function) of a trait, that is, its contribution to the organism's fitness. For example, to explain why certain fish switch from asexual to sexual reproduction under environmental stress, an adaptationist will examine the advantage that accrues to a fish who uses this switching strategy over one who doesn't, in terms of increased reproductive success. Though adaptationist reasoning has its

[10] See for example Valone (2006) or Dall *et al.* (2005).

limits, and must be used carefully, the fact is that many behavioural, physiological, and morphological traits have been successfully explained in this way.

Biologists engaged in adaptationist explanation often treat an evolved organism as an agent with a goal or objective. Thus an organism's 'ultimate' goal is said to be maximizing its fitness; to achieve this goal it needs to pursue intermediate goals such as survival, finding food, and attracting mates, to which its various traits make distinct contributions. For example, Roff (1992), in a textbook on life-history theory, writes: 'the primary goal of any organism is to reproduce . . . the first 'decision' it must make . . . is when to start reproducing' (p. 2), while West and Gardner (2013), in a discussion of social evolution, describe maximizing inclusive fitness as the 'objective' of an organism's social behaviour. More generally, Grafen (2014a) argues that natural selection will lead an organism to behave like a 'maximizing agent' trying to maximize an 'objective function'; he regards this as a formalization of Darwin's argument that selection leads to the appearance of design.

A related agential idiom that often arises in this context is that of 'interests'. Thus the male and female birds in a breeding pair have overlapping but non-identical interests; a worker honey bee has a greater genetic interest in the queen's offspring than in those of other workers, while a horizontally transmitted virus has no long-term interest in its host's survival. Such locutions go hand in hand with treating an evolved organism as an agent with a goal. For an organism can be said to have interests only in so far as some circumstance benefits or harms it, that is, promotes or detracts from its goal.

This third rationale for treating organisms as agents should be distinguished from the previous two, for it is explicitly evolutionary and thus pertains to ultimate not proximate explanation. Moreover, it has nothing to do with behavioural flexibility or goal-directed activity per se. To illustrate, consider a sessile organism such as a cactus that is well-adapted to its environment. A cactus's spines have the function of deterring herbivores, which contributes to its survival and reproduction. But the cactus lacks behavioural flexibility—it still grows spines in a greenhouse where herbivores are absent, and deterring herbivores is not plausibly regarded as an activity of the cactus—it is not something that the cactus *does*. (Thus we would not describe a cactus as 'trying' to deter herbivores by growing spines.) So although the cactus certainly has traits that further its ultimate goal, it lacks the other attributes of agency.

In other cases, though, the third rationale dovetails neatly with the first two. Where organisms do exhibit goal-directed activity and flexible behaviour, these do usually contribute to their ultimate goal of survival and reproduction. For example, a squirrel's nut-storing behaviour is both goal-directed and flexible, and the goal towards which it is directed—provisioning for the winter—obviously benefits the squirrel's survival prospects and thus its fitness. The same is true of most evolved animal behaviour. Note also that adaptive behaviour typically depends on morphological adaptations that are not themselves flexible, for example, the squirrel's ability to crack nuts depends on its oversized incisors and its powerful jaws. So although many organismic adaptations are hard-wired, hence in no sense 'chosen' by the organism to further their goal, they are typically necessary for the organism to display flexible, goal-directed behaviour. Thus the three rationales for treating organisms as agents, though logically distinct, are often related empirically.

The cactus example might lead one to wonder whether the third, evolutionary rationale is itself sufficient for talk of agency to be appropriate. After all, real agents make choices and perform actions; so surely an evolved organism is only usefully regarded as agent-like in so far as it exhibits flexible behaviour? There are two possible ways to go here. The first is to concede the objection and restrict talk of agency, in an evolutionary context, to organisms with sufficient behavioural flexibility that the language of rational choice applies, that is, the organism must be able to adjust its behaviour in pursuit of its (ultimate) goal; so cacti are ruled out. The second option is to insist that the third rationale for treating organisms as agents is self-standing, and independent of the other two. After all, any evolved organism can sensibly be treated as having the goal of surviving and reproducing; and its traits can be evaluated in terms of how well they contribute to this goal. Thus talk of the organism's 'interests'—a paradigmatic agential idiom—makes good sense irrespective of how plastic its behaviour is.

I incline towards the second option. For one thing, on the first option it is unclear how much behavioural plasticity counts as sufficient. A cactus may be out; but what about a climbing plant which tries to grow up a glass rod but on failing unwinds and searches elsewhere for a suitable support?[11] Such plant behaviour is quite flexible, though less so than that of a foraging bird; but where do we draw the line? Moreover, a sessile organism with purely hard-wired traits still has to *develop*; and ontogeny is of course a dynamic, goal-directed process. So even when the second rationale (behavioural flexibility) does not apply, the first rationale (goal-directedness) still will, and this is reflected in the agential idioms that biologists use to express adaptationist hypotheses. We do not describe a cactus as 'trying' to deter herbivores, but it is unexceptionable to say that the cactus develops (or grows) spines in order to deter herbivores. So even a sessile organism 'moves' towards its goal in this ontogenetic sense, which lessens the oddity of treating it as an agent.

Nonetheless, it might still be wondered whether talk of agency is doing any real work in an evolutionary context, if used this broadly. Surely we can construe adaptationist explanations in a way that makes no reference to agents, goals, or objectives, simply by saying 'the function of the trait is such-and-such', where this is understood in the usual way, as telling us why the trait evolved or was maintained in the population? What does agential talk achieve that could not equally be achieved by talk of adaptive significance, or function? I address this challenge in the next section.

[11] This example comes from a book on 'plant intelligence', a notion which the author understands literally; see Trewavas (2015), p. 271.

Although I see the third rationale for the 'organism as agent' concept as self-standing, and applicable to adaptive traits of all sorts, it becomes particularly interesting in the context of evolved behaviour. In this context the adaptationist penchant for agential thinking often takes a particular form, in which the behaviour's evolutionary function is treated as if it were the organism's reason for performing the behaviour, and the adaptationist explanation is re-cast in an intentional idiom. Why do swallows migrate? Because they want to escape the cold. Why do female rats kill their congenitally malformed offspring soon after birth? Because they know that the offspring will not survive and don't want to waste resources on them. Why do worker honey bees eat the eggs laid by fellow workers? Because they prefer that the offspring of the queen be reared instead. In this way the language of instrumental rationality is used to describe and theorize about evolved behaviour: the organism is treated as an agent who acts for reasons and pursues a goal. Thus when agential thinking (type 1) is applied to behaviour, this yields what I call the 'organism-as-rational-agent' heuristic.

Expressing adaptationist explanations of behaviour in this way is potentially misleading, as it can lead to confusion of ultimate with proximate cause (Scott-Phillips *et al.* 2011). The function of a bee's egg-eating behaviour is indeed to enable more of the queen's eggs to be reared, but the proximate explanation of the behaviour is hormonal, not psychological. However, as long as we are alert to this danger, and are clear that intentional idioms in this context pick out the behaviour's evolutionary function not its proximate cause, they do no harm. Indeed, I argue below that they are actually useful. Note also that this evolutionary use of intentional idioms is distinct from their use in fields such as cognitive ethology, where the focus is on proximate explanation. The honey bee's nervous system is too simple, and its behaviour too rigid, for a proximate explanation in intentional-psychological terms to be plausible; but this does not prevent us construing the evolutionary explanation of its behaviour, metaphorically, in terms of what the bee prefers or wants.

To summarize: we have found three distinct rationales for agential thinking (type 1), that is for treating organisms as agent-like: goal-directedness, behavioural flexibility, and possessing traits that are adaptations. The three rationales are independent, and are associated with different agential and intentional usages. However, they dovetail in some cases, in particular where complex animal behaviour is concerned. All three rationales are defensible, and all are found in actual biological practice. In the remainder of this chapter and the next, it is the evolutionary rationale for agential thinking that will occupy centre-stage.

1.6 Unity-of-purpose

Effective agency . . . requires a unity-of-purpose both at a time, in order that we may eliminate conflict among our motives and do one thing rather than another, and over time, because many of the things we do form part of longer-term projects and make sense only in the light of these projects and plans. Kennett and Matthews (2003), p. 307

Let us return to the question: what does talk of agency achieve, in an evolutionary context, that talk of function does not? It is widely accepted that evolved traits, behavioural and non-behavioural, can be ascribed functions, often in a fairly determinate way; this is a commonplace in evolutionary biology. Thus the function of a crab's exoskeleton is to protect its innards, of a peacock's tail to attract mates, and of a meerkat's warning cry to alert its companions. By 'function' here I mean adaptive significance, that is, contribution of the trait to fitness which explains why it evolved.[12] The issue is whether introducing talk of agents with objectives, goals, and interests adds anything to this.

I think that it does, for the following key reason. Functional talk applies to *traits*, but agential talk applies to *organisms* (or genes or groups, in some cases). We talk about the function of a particular trait; but it is the whole organism, not its traits, that has a goal or objective, or prefers one thing to another, or has interests that can be hindered or advanced. Similarly, when intentional idioms are used in an evolutionary context, the subject of the intentional attribution is the whole organism, not one of its traits. The meerkat's warning behaviour has a function, but it is the meerkat that sees the danger and wants to warn its companions.

Why does this matter? Primarily because it highlights an implicit theoretical commitment of agential thinking (type 1) in biology, namely that the entity which is treated as an agent—assumed here to be an individual organism—possesses a 'unity-of-purpose', in that its different traits have evolved because of their contributions to a *single* overall goal: enhancing the organism's fitness. Where this unity does not obtain, the organism cannot be regarded as agent-like, and treating it as such will impede, not facilitate, understanding of its features in adaptationist terms.

To appreciate this point, note that unity-of-purpose is fundamental to human agency. This unity has two components. Firstly, a person's goals (or intentions) should cohere with each another in the sense of being mutually reinforcing, or at least not clearly inconsistent; secondly, their actions should tend to further their goals, that is, they should be instrumentally rational. Minor deviations from this unitary ideal are common, but if they are too many, or too great, it becomes impossible to treat the person as a unified agent, and to rationalize their actions in terms of their goals. This is a familiar theme in the philosophy of mind and action.[13] In the biological case, an analogous unity-of-purpose is necessary in order for an evolved organism to be treated as agent-like, and is presupposed when agential idioms ('interests', 'goals')

[12] Here I assume for simplicity that a trait's current utility is the same as the reason for which it originally evolved. Where this is not true, finer distinctions are needed, as Tinbergen (1963) famously argued; see Bateson and Laland (2013) for a good recent discussion.

[13] That a subject counts as an intentional agent only in so far as their actions and beliefs are rational is emphasized by Davidson (2001a, b) and Dennett (1987). That agents are rationally required to have consistent intentions, and to exhibit means-end coherence, is emphasized by Bratman (1987). Similarly, Rovane (1998) argues that agents are required to exhibit a 'rational unity', while Korsgaard (1989) describes 'unity of agency' as stemming from 'the raw necessity of eliminating conflict among your various motives', and as 'implicit in the standpoint from which you deliberate and choose' (pp. 110–11).

and intentional idioms ('wants', 'tries') are applied to it in an evolutionary context. By contrast the functional idiom, since it applies on a trait-by-trait basis, involves no such presupposition.

To illustrate, consider a case where the required unity-of-purpose does not obtain: the phenomenon of cytoplasmic male sterility in flowering plants. This phenomenon, puzzling at first sight, arises from conflict between the genes in the nucleus and the mitochondria over gamete production. Flowering plants are usually hermaphroditic, making both pollen and ova, which is what nuclear genes 'prefer', that is, are selected to bring about. However, mitochondrial genes are only transmitted maternally, so from their viewpoint producing pollen is a waste. They thus try, and sometimes succeed, to cause their host plant to abandon pollen production altogether. Now the male sterility trait certainly has a function, namely facilitating the transmission of mitochondrial genes to the next generation. But we cannot sensibly think of the plant itself as having this as its goal, for it is detrimental to it, preventing it from selfing, and the plant has other traits which conflict with this goal.

To make the organismic disunity here more apparent, suppose that the sterile plant does develop stamen—specialized organs for producing pollen—but that no pollen is actually made. (This is one form that cytoplasmic male sterility actually takes.) Like all specialized organs, stamen are energetically costly to produce. We can give a functional explanation for why the plant develops stamen: they are there to make pollen and thus facilitate sexual reproduction. We can also give a functional explanation for why no pollen is made: mitochondrial genes gain by suppressing it. But we cannot take an agential perspective and explain both of these traits as contributions to an overall 'goal' of the plant. For the traits have mutually antagonistic functions, pulling the plant in different directions. The plant is akin to a victim of split-personality disorder, who both wants and does not want a certain outcome, rather than to an agent with a rational unity-of-purpose.

Examples of this sort could easily be multiplied, for a certain amount of intra-organismic conflict is relatively common in modern organisms. To take one further example, in the fruitfly *Drosophila pseudoobscura*, males that carry a particular X-chromosome variant produce no Y-bearing sperm at all, as a result of 'sperm killer' genes on the X-chromosome which disrupt spermatogenesis (Burt and Trivers 2006). As a result, far fewer viable sperm are produced than in normal males. This trait—failure to produce Y-bearing sperm—evolved not because it benefits the organism, which it does not, but rather because it benefits the X-chromosome itself (or the genes on it). So again, the fruitfly exhibits a (partial) disunity-of-purpose. Some of its traits, for example, its mating behaviour, pull in the direction of maximizing its reproductive success, while others pull in a different direction. If a biologist studying fruitfly spermatogenesis treats the fruitfly as a unified agent, they will not understand what they see.

A similar failure of unity occurs in cases of parasitic manipulation. Consider for example an ant infected by the liver fluke parasite *Dicrocoelium dendriticum*. This parasite induces a change in the ant's behaviour, causing it to climb to the top of a blade of grass every evening and stay there, clamped to the tip with its mandible. This increases the chance that the ant will be ingested by a sheep, which is what the parasite needs in order to complete its life-cycle. Again, an infected ant illustrates a partial disunity-of-purpose. Some of its traits, for example, its foraging behaviour, further its own goal of survival and reproduction, while others, for example, its nightly ascent of a blade of grass, detract from that goal. If a biologist tries to treat the ant as a unified agent they will not understand what they see.

The cytoplasmic male sterility and sperm-killer examples illustrate how unresolved intra-genomic conflict—that is, conflict between the genes within a single organism—undermines the unity-of-purpose necessary for the organism to be treated as an agent in pursuit of goal. However, it may still be possible to apply agential thinking in a different way, by treating *genes* as the agents and regarding the organism as a body inhabited by multiple agents with divergent goals. The subjects of agential/intentional attributions, that is, the entities with interests and goals which they 'try' to achieve, would then be genes rather than organisms. This possibility is explored in the next chapter.

The contrast between agential/intentional attributions to organisms and functional attributions to traits is reminiscent of a traditional contrast in the philosophy of psychology, between personal and sub-personal attributions.[14] Folk-psychological notions such as believing, desiring, and reasoning are personal-level: it is whole persons that occupy these states or undergo these processes. By contrast, the computational processes described by cognitive psychology are sub-personal; they are carried out not by whole persons but by parts of their brains. Thus the cerebral cortex processes visual information, but the whole person sees the approaching car and moves out of the way. Essentially, we have here an evolutionary analogue of this distinction. It is the parts of an organism, that is, its traits, that have Darwinian functions, but it is the whole organism that has aims, goals, and preferences. Moving from the former sort of attribution to the latter is only possible in so far as the organism exhibits a unity-of-purpose in the sense described above.

Another way to see the point is this. Ordinarily a statement of the form 'the function of the organism's trait is *x*' can be transposed into the form 'the organism has the trait in order to achieve *x*', in which the organism itself is the grammatical subject. For example, that the function of a bird's display is to attract mates can equally be expressed by saying that the bird performs the display in order to attract mates. However, this transposition is less innocent than it seems, for it cannot be applied across the board when an organism's traits have antagonistic functions. The function of ceasing pollen production in flowering plants is to promote mitochondrial

[14] This distinction comes from Dennett's 1969 book *Content and Consciousness* and is widely invoked in the philosophical literature; see Drayson (2014) and Frankish (2009) for discussion.

transmission, but the plant does not cease to make pollen in order to achieve this. Indeed in a very real sense, ceasing to make pollen is not something that the plant *does* but something that is done to it, and similarly for impeded spermatogenesis in the fruitfly.

The general moral is this. To treat an evolved organism as agent-like requires that we can treat the organism's various traits as instruments for achieving sub-goals—finding food, keeping warm, producing gametes, mating—which contribute to a single overarching goal, namely enhancing the organism's fitness. Empirically, this requires that the genes coding for the traits have identical evolutionary interests, so that the traits evolve functions which are complementary rather than antagonistic. Otherwise, then although each trait considered individually can be given a functional explanation, the organism cannot be treated as a unified agent.

This point is a corollary of a widely accepted evolutionary principle, namely that internal conflict tends to undermine the integrity of a larger unit. Thus multi-celled organisms have evolved numerous mechanisms for suppressing conflict among their constituent genes and cells, including fair meiosis, uniparental inheritance of organelles, and programmed cell death (Maynard Smith and Szathmáry 1995, Bourke 2011). It is because these mechanisms usually work well that organisms are as cohesive and integrated as they are, though some residual conflicts remain (Queller 1997). The same principle applies at higher levels: those insect colonies that come closest to being 'superorganisms' are precisely those in which within-colony conflict is effectively suppressed (Ratnieks *et al.* 2006, Bourke 2011). This biological principle is an empirical one, but it has a conceptual counterpart. It is only because an organism's constituent traits typically contribute towards a single goal that the organism can be treated as agent-like.

Since the unity-of-purpose requirement is not always satisfied, this might be regarded as a limitation of agential thinking (as applied to organisms). In a way this is so, but it also shows that the organism-as-agent concept has a genuine rationale and is not mere idle metaphor. For most of the time, evolved organisms manage to suppress intra-genomic conflict, thanks to a suite of traits specialized for this, so the requisite organismic unity does obtain, at least to a high degree of approximation. Thus there is a real pattern in nature that is captured by treating the organism as an agent, which the functional idiom does not capture, and which gives agential thinking in evolutionary biology both its power and its point.

For human agents, we identified two components to unity-of-purpose: consistency of an agent's goals with each other, and a suitable fit between the agent's goals and their actions. The evolutionary analogue of the former is having traits that all contribute to the organism's ultimate goal—this is what fails in the flowering plant and fruitfly examples above. Is there also an evolutionary analogue of the latter? The answer is yes, but only for organisms that exhibit behavioural flexibility. The analogue is simply that the organism's evolved behaviour, and thus its within-lifetime choices, should contribute to its ultimate goal.

To illustrate, recall that adaptationist explanations of behaviour are often couched in intentional terms, as discussed previously. ('The rat kills malformed pups because it knows they won't survive and doesn't want to waste resources.') Importantly, this use of intentional language is only possible when the organisms' evolved behaviour fits suitably with their (ultimate) goal, just as intentional explanation of a human action is only possible when the action fits suitably with the agent's (proximate) goal. To see this, imagine that rat infanticide worked slightly differently. Suppose that a gene in an adult female somehow caused it to kill any pups in its litter that do not contain the gene. (This is not as far-fetched as it may sound.[15]) Such a gene could easily spread by natural selection. If so, then, although an adaptationist or functional explanation of the rat's behaviour could still be given, it could not be couched in terms of what the rat 'wants'. For the behaviour would harm, not further, the rat's evolutionary interests. The rat would be analogous to a person whose actions fail to cohere suitably with their goals, and who thus lacks (the second component of) unity-of-purpose.

In section 1.5.3, we considered the objection that talk of agency is inappropriate if applied to organisms that lack behavioural flexibility. A desert cactus is well-adapted to its environment but not very agent-like, the objection went. Our discussion of unity-of-purpose provides a further reason to reject this objection. For although the second component of unity-of-purpose does not apply to cacti, the first component does. We can ask of any organism, including a plant, whether its evolved traits are complementary or antagonistic, as the example of cytoplasmic male sterility illustrates, and only in so far as they are complementary can the organism be treated as agent-like. Thus there is a genuine point to treating an evolved organism as an agent with a goal, that does not depend whether the organism's means of achieving the goal are behavioural or not.

Our discussion of organismic unity should be sharply distinguished from an argument of Trivers (2009), which postulates a causal link between intra-genomic conflict and internal conflict in the human psyche. Trivers focuses on genomic imprinting, in which a gene has different phenotypic effects depending on whether it is paternally or maternally inherited. This leads to intra-genomic conflict; for if a gene in an organism is paternally inherited then it has no genetic interest in the future reproduction of the organism's mother, while genes at other loci do. Trivers suggests that this will have psychological consequences: 'we literally have a paternal self and a maternal self and they are often in conflict' (2009, p. 163). Similarly, Haig (2006, 2008) suggests that intra-genomic conflict may lead to a 'divided self', in which the human mind is inhabited by multiple personae with different agendas.

The Trivers/Haig hypothesis is interesting though speculative. I take no stand on the matter here.[16] I do not claim a direct connection between evolutionary

[15] David Haig's kinship theory of genomic imprinting has uncovered phenomena of exactly this sort; see Haig (2002, 2004).

[16] See Spurrett (2016) for a good critique.

unity-of-purpose and unity-of-purpose in human agents (or their absence). Rather, my point is that the attribution of goals, purposes, and intentions to a subject only makes sense if the subject exhibits sufficient unity, that is, it is an undivided self, or close enough. So to treat an evolved organism as akin to an agent trying to achieve a goal, for the purposes of adaptationist theorizing, requires that the organism's traits do not have mutually antagonistic functions; and empirically, this requires the absence of intra-genomic conflict or nearly enough. This is a claim about the presuppositions of a particular psychologically derived idiom that we apply, usually metaphorically, to evolved organisms; it is not a claim about the evolutionary roots of psychological unity or disunity in humans.

We have defended the unity-of-purpose requirement by essentially philosophical means. We noted that in general, attributions of agency presuppose that an agent's intentions are consistent and that their actions fit suitably with their goals. So when an evolved organism is treated as agent-like, in a biological context, an analogous unity is presupposed, namely that the organism's traits contribute to a single overall goal. Interestingly, a quite different defence of unity-of-purpose emerges from Grafen's work on what he calls the 'individual-as-maximizing-agent analogy' in biology (Grafen 2002, 2006, 2014a). Grafen offers a precise characterization of when individual organisms will behave like maximizing agents, from which it follows deductively that intra-genomic conflict must be absent. This is strikingly consonant with our conclusion above, since we have seen that intra-genomic conflict, unless suppressed, renders the agential idiom inapplicable. This connection, and the underlying logic of Grafen's argument, are examined in section 4.2.

Finally, a minor irony deserves note. One of Gould and Lewontin's objections to adaptationism in their famous *Spandrels* paper was that it illegitimately 'atomizes' organisms into traits then seeks a separate adaptive explanation for each, ignoring the fact that organisms are 'integrated entities' (Gould and Lewontin 1979). There may be some truth to this charge. However, those adaptationists who employ agential thinking, and express their adaptationist hypotheses via agential and intentional attributions to evolved organisms, are actually *presuming* that organisms are integrated units. For such attributions only apply when organisms exhibit the required unity-of-purpose, in the sense of having traits which contribute to a single overall goal, as described above. Therefore one strand of adaptationist thinking, at least, does not ignore organismic integration so much as presuppose it.

1.7 Agents, Goals, and Interests

Biologists generally employ agential terminology—the language of goals, interests, and preferences—in a fairly casual way. This is not necessarily a problem; but it is useful to seek a more precise way of expressing the idea that a biological entity, of any sort, is agent-like. One way to do this is to import a formalism from rational choice theory. This allows us to make agential talk precise, and where multiple agents are concerned, to express the idea that their interests may conflict. Absence of such conflict is intimately related to the unity-of-purpose constraint described in the previous section.

In the simplest form of rational choice theory, an agent is faced with a finite set of options X, which are pairwise exclusive. For example, X could be a set of alternative holiday destinations. The agent has a *preference order* over the options, denoted by R. The formula aRb means that the agent (weakly) prefers a to b, where $a, b \in X$; this is usually interpreted to mean that the agent never chooses b over a when both options are available. Formally, R is a binary relation on the set X, that is, a subset of $X \times X$. From R, it is straightforward to define the agent's strict preference P and indifference I; thus aPb if and only if aRb but not bRa; while aIb if and only if aRb and bRa.

Since the agent is rational, their preference order R is assumed to be both complete and transitive. Completeness means that either aRb or bRa for all $a, b \in X$, transitivity that aRb and bRc implies aRc. These conditions imply that the agent's preferences can be represented by a real-valued utility function u on X, that is, $u(a) \geq u(b)$ if and only if aRb, for all $a, b \in X$. The utility function u thus embodies the agent's goal, in the sense that the agent behaves as if they are trying to maximize it. Note that the utility function in this case is purely ordinal, that is, any monotone transformation of u will also represent the agent's preferences.

To apply this formalism to biology, consider the simple case in which the agent is an individual organism whose fitness depends only on its own phenotype (in a given environment). The set X may then be taken as a set of alternative phenotypes; these could be behavioural, physiological, or morphological. For example, X might contain alternative life-history schedules, each specifying the age at which reproduction should begin. There is a fitness function on X; the organism's 'goal' is to choose the phenotype that maximizes it. It may be that the organism performs an actual choice between the options within its lifetime; alternatively, its phenotype may be hard-wired. In the latter case, talk of 'choice' means simply that it is as if the organism had chosen its phenotype. The organism's fitness function corresponds to the utility function u. It is then straightforward to define the organism's binary preferences on X—it always prefers the phenotype that brings higher fitness. The resulting preference relation will necessarily be complete and transitive. This allows precise meaning to be given to talk of an organism's 'interests', 'goals', and 'preferences' in an evolutionary context.

Note that in standard rational choice theory, an agent's utility function is derived from their preference order, which itself summarizes their choice behaviour. (The motivation for this 'revealed preference' approach, as it is called, is that an agent's choices are observable while their utility function is not.) By contrast, in the biological interpretation of the formalism, the organism's fitness function comes first, and is used to define their preference ordering. (Though the fitness function contains more than merely ordinal information; see section 6.6.) Thus the properties of transitivity and completeness do not have to be postulated, but are derived.

Agential thinking as applied to genes and groups may be expressed similarly, by suitable choice of *X*. If the agent is an insect colony, for example, and the trait of interest is allocation of workers to different tasks, such as foraging, feeding the brood, and colony defence, then *X* could be defined as a set of alternative allocations, each of which confers a particular reproductive fitness on the colony as a whole. This allows us to capture talk of the colony's 'preferred' allocation of workers to each task, and of its 'goal' in moving workers from one task to another in response to external perturbations. If the agent is a gene at a particular locus in an organism, *X* might be defined as the set of alternative genetic compositions of the organism's successful gametes. This allows us to capture the familiar idea that the gene 'wants' to leave as many copies of itself as possible in its host organism's gametes, and thus 'tries' to bias meiosis in its own favour.

Another biological idiom that the rational choice formalism can capture is talk of 'interests', in particular conflicts of interest between different agents. Consider two agents with preferences on the set *X*. If their preference orders are identical, then the agents are said to have 'common interests'; this means that their goals are aligned, so conflict will not arise. Conversely, if one agent's preference order is the inverse of the other's then there is a 'pure conflict of interest'; in effect, this means that the agents are playing a zero-sum game. Between these two extremes, the two agents' preferences may overlap partly but not completely; interests are then said to be 'mixed'. This terminology is standard in game theory.

To see this at work, consider the conflict between workers and queen over the sex-ratio in a typical social insect colony. A once-mated queen will favour an equal colony sex-ratio, whereas a worker will prefer a sex-ratio biased 3:1 in favour of females, since workers are more closely related to females than males (Trivers and Hare 1976).[17] If *X* is the set of all possible colony sex-ratios, then worker and queen will have different preferences over *X*, so their goals will conflict. This is a case of mixed interests, for the workers' and queen's binary preferences agree over some pairs in *X* (e.g. 3F:1M versus 4F:1M), but disagree over others. Other conflicts of interest in biology, for example, between nuclear and mitochondrial genes in a single organism, or between males and females in a breeding pair, can be captured similiarly.

For a biological entity to evolve adaptations, and thus be agent-like, a mechanism is usually needed to align the interests of its constituent parts by suppressing conflict, as discussed in section 1.6. Thus multi-celled organisms and certain social insect colonies have evolved mechanisms, such as fair meiosis, apoptosis or programmed cell death, and worker policing, for regulating the activities of their constituent genes, cells and insects, ensuring that they work for the common good (Queller and Strassmann 2009, Bourke 2011). In rational choice terms, such mechanisms work by preventing one agent or more from being able to achieve their most preferred outcome, that is,

[17] This presumes that the species in question is haplodiploid, that is, males develop from unfertilized eggs so contain only one set of chromosomes.

by restricting their feasible choices to a subset of the options. This has the effect of converting the situation into one of common interest.

To illustrate, consider fair meiosis. The genes at different loci within a diploid organism have interests that potentially conflict—each prefers that the organism's successful gametes contain as many copies of itself as possible. But if meiosis is constrained to be fair, then any gene can only influence the *number* of successful gametes that are produced, not the proportion of them in which it is found. So on the feasible set, all genes have identical preferences—that the organism leave as many successful gametes as possible—and they all work together for this end. Similarly, worker policing of reproduction in honey-bee colonies, in which workers eat the eggs laid by other workers, ensures that no worker is able to 'choose' the outcome it most prefers—having offspring of its own; thus a worker's choices are restricted to actions that affect the reproductive output of the queen. On this restricted set, interests are aligned—each worker prefers that the queen maximize her reproductive success.

Recall the unity-of-purpose constraint: for a biological entity to be treated as an agent, its evolved traits must have complementary rather than antagonistic functions; otherwise the entity cannot sensibly be regarded as 'trying' to achieve a goal by means of its traits. This is a conceptual point; empirically, it requires the absence of internal conflict, and thus mechanisms for suppressing or minimizing conflict. Therefore, the foregoing characterization of how suppression mechanisms work—by restricting the feasible set and thus creating a situation of common interest—can be regarded as an indirect characterization of how biological unity-of-purpose is often achieved.

The formalism of rational choice theory is a way of capturing what biologists mean when they apply agential thinking to organisms, groups, and genes, and attribute to them goals, interests, and preferences. It does not in itself justify agential thinking; nor does it help determine whether the relevant agent in any particular case is an organism, group, or gene; nor does it tell us how the agent's fitness function should be defined. These are all substantial issues; the point of the formalism is simply to provide an abstract language for making agential thinking more precise, and in particular for expressing the idea that agents' interests may be in common or in conflict.

It is worth explaining briefly why the choice of fitness function, in terms of which an agent's preferences and interests can be defined, may be less straightforward than it seems. Take the case where the agent is an individual organism. A standard definition of individual fitness is 'lifetime reproductive success', that is, total number of surviving offspring. However this definition is not always appropriate, for the type with the highest value of this quantity will not always be at a selective advantage. If the population is expanding, selection may favour individuals who reproduce early even if their total lifetime reproduction is reduced, as they will make a greater genetic contribution to future generations (Charlesworth 2008). The relevant fitness measure is then Fisher's 'reproductive value', which takes account of this complication (Fisher 1930). Or to take another example, if individuals engage in social interaction, then a suitable measure of fitness needs to take account of an individual's contribution to its relatives' reproductive success as well as its own. The point of Hamilton's 'inclusive fitness' concept is precisely to supply such a measure; see chapter 5. The same point applies to other agents too. In the literature on group selection, for example, there is an ongoing discussion about how a group's fitness should be defined—in particular, whether it should be equated with average individual fitness or defined independently, for example, in terms of group persistence and reproduction.[18] The 'right' definition depends partly on the details of the evolutionary model, and partly on what question we are trying to answer.

These examples highlight some of the trickiness of the fitness concept in biology, but that is not what I want to emphasize here. Rather the point is this. We have seen that the language of goals, interests, and preferences, as used in evolutionary analysis, is derivative from the fitness function and may be defined precisely in terms of it. Therefore whenever such language is used, implicitly this presumes that an appropriate choice of fitness function has been made. This choice is not a subjective matter, for the fitness concept plays a specific theoretical role—the fittest type is meant to be the one that will evolve by natural selection.[19] The appropriate fitness definition thus depends on the details of the underlying evolutionary process, or model, and will differ from case to case.

According to Steve Frank (2003), 'discussing "interests" in arguments about how natural selection operates can be misleading' (p. 700). The context of Frank's remark is an exploration of how alignment of interests among lower-level units (e.g. cells) permits a higher-level unit (e.g. a multi-celled organism) to evolve adaptations. Frank regards this as a key principle of social evolution and provides examples of the principle at work in diverse social systems. So he is not opposed to all talk of 'interests' in evolutionary biology. Rather, his point is to caution against the inference from a circumstance being in some agent's interests to the conclusion that that circumstance will evolve by natural selection. Frank is quite right that this inference is suspect, particularly in the context of multiple levels and/or timescales. It is perfectly conceivable that a given outcome might benefit an organism or group were it to evolve, but short-term selective pressures may mitigate against it. For example, sexual reproduction may be in the interests of a species, as the resulting genetic heterogeneity may increase the species' survival prospects if the environment changes; but this is unlikely to be the reason why sexual reproduction originally evolved.

This point is important, and a useful corrective against free-wheeling use of agential thinking in biology. However, it remains the case that talk of interests is commonplace in biology and is often relatively unproblematic. Many biological phenomena can

[18] See Damuth and Heisler (1988), Okasha (2006), and Gardner and Grafen (2009) on this issue.

[19] This is an oversimplification, since it is not always clear what 'will evolve by natural selection' means. If there is stochasticity, for example, does it mean that the type's *expected* frequency will increase? Does it refer to short-term or long-term increase? Does it mean that the type's probability of fixation is higher than that of competing types? Despite these complications, the main point in the text—that the appropriate choice of fitness measure is determined by what we want the fitness concept to do—is still valid.

legitimately be explained by showing how they further an agent's interests, or promote their goal; it is often possible to express adaptationist explanations in such terms, and biologists often do. This section has shown how such talk can be formally captured in the language of rational choice, taking fitness as a primitive.

1.8 Dennett Reconsidered

Here I briefly revisit Dennett's idea that adaptationist theorizing in biology involves 'reading mother nature's mind', in the light of the foregoing discussion. I think Dennett has an important point, but that it is better captured by agential thinking of the first type—in which the agent is the individual organism rather than 'mother nature'.

Dennett's starting point is the apparent parallel between intentional explanations in psychology and adaptationist explanations in evolutionary biology (Dennett 1983, 1987). When we give a folk-psychological explanation of an agent's action, or 'adopt the intentional stance', we explain the action in terms of the agent's reasons for acting: their beliefs, desires, and goals. Such explanations work by showing that the action serves the agent's goal, so is instrumentally rational. When a biologist gives an adaptationist explanation of an organismic trait, for example, an evolved behaviour, they explain how the trait serves the organism's biological goal, that is, enhances its fitness. Dennett thus argues that rationality in folk psychology corresponds to optimality in evolutionary biology: they are both presuppositions of the respective modes of explanation.

I agree with Dennett that this parallel is deep and instructive. For notice that both modes of explanation have an irreducibly *normative* dimension. In adaptationist explanation, a trait is identified as an optimal or near-optimal fit to the environment, given the organism's biological goal of survival and reproduction (given an assumption about the alternative possible traits). In psychological explanation, an agent's action is identified as an optimal thing for the agent to do, given their beliefs and desires (given an assumption about the alternative possible actions). In both cases we begin with a conception of how things *should* be in order that a particular end be achieved, and try to show that how things actually are is a close match. The same is not true of scientific explanation in general. Newton's gravitational theory explained why planets move in elliptical orbits, but not by showing that they should do so. When Mercury's observed orbit was not as predicted, this was a puzzle that demanded scientific attention, but not something that was inappropriate, irrational, or maladaptive.

Dennett's parallel is bolstered by the well-known philosophical distinction between two ways in which a phenomenon can be explained, or rendered intelligible.[20]

[20] This distinction is emphasized by Davidson (2001a, b) and McDowell (1994), though neither of these authors would agree that adaptationist explanation is of the rationalizing sort.

The first involves subsumption under law: showing that the phenomenon had to happen as a matter of nomological necessity. The second involves rationalization: showing that the phenomenon makes sense, or is appropriate to a particular end. Psychological explanation is of the latter sort, and arguably, so too is adaptationist explanation. The latter explains a trait's presence by showing that, from the perspective of the organism, it makes sense to have it. Moreover, there is no subsumption under laws. As has often been observed, adaptationist reasoning is typically post-hoc, explaining features after they have been discovered, and has only modest predictive power. Whether this is problematic is another question; the point is simply that a typical adaptationist explanation, for example, of why swallows migrate in winter, involves no overt laws of nature. Rather, the trait is rendered intelligible by showing that it serves the organism's purposes, that is, is fitness-enhancing. The type of intelligibility here is more akin to that we get from understanding why someone has acted the way they did, than from seeing that a phenomenon was mandated by natural law.[21]

To this point I am in agreement with Dennett. However, Dennett's next move is to bring in 'mother nature', via the following line of reasoning. Evolutionary thinking is about finding the (ultimate) reasons why organisms display the traits that they do. For example, the reason why a fledgling cuckoo chick, immediately on hatching, tries to roll other eggs out of the nest is to secure more resources for itself. However, the cuckoo chick's behaviour is purely instinctive. It is 'mother nature', that is, natural selection, who has worked out that this is a good strategy for the cuckoo, by trial and error over generations. So adaptationist biology, as Dennett conceives it, is an essentially hermeneutic enterprise. It involves trying to read mother nature's mind or to decipher her intentions, for example, to discover why she equipped the cuckoo with the instincts that it has.

I find Dennett's appeal to 'mother nature' here under-motivated, and not necessary to capture the parallel between adaptationist and psychological explanation that I take to be his main insight. On the contrary, that parallel fits better with treating the evolved organism, not natural selection, as agent-like. For what drives the parallel is precisely the fact that both explanatory modes involve reference to some circumstance furthering an agent's interests, or benefiting them, or helping them achieve their goal. But in the adaptationist case, the agent with the interests, whose benefit is in question, is the evolved organism, not mother nature. It is not natural selection that has goals and interests (such as 'improving the population') in terms of which we can rationalize an organism's evolved traits, but the organism itself. Indicative of this is that when biologists couch their adaptationist explanations of behaviour in the intentional idiom, the subject of the intentional attribution is the organism, not mother nature. In short, I agree with Dennett that adaptationist explanation has a hermeneutic quality; but I think this is better construed not as a search for mother nature's reasons, but

[21] This is not to say that there are no general laws or principles in evolutionary biology, but rather just that adaptationist explanation does not fit the 'covering law' model of explanation very well.

rather as an attempt to understand how an organism's evolved traits help further its biological interests.

A related worry about Dennett's position is that his invocation of 'mother nature' sometimes comes close to the idea, criticized previously, that the process of natural selection can legitimately be analogized to a fictitious agent's choice. In Dennett's writings, the claim that adaptationist reasoning involves reading mother nature's mind is often accompanied by an agential description of the process of natural selection itself: it involves mother nature's 'choosing' between phenotypes for her own 'reasons' (1987, p. 259). These ideas are separable—the former is a thesis about adaptationist explanation, the latter about the process of natural selection—but Dennett appears to move between them. We saw in section 1.4.1 that the metaphor of agential choice is a problematic way of thinking about natural selection, as it only really makes sense where the selective environment is constant, and can easily mislead us into thinking that natural selection is inherently directional, mimics a goal-directed process, or leads to maximization of some quantity or other. I do not claim that Dennett himself commits these errors, but the 'mother nature' metaphor certainly invites them.

Finally, I want to comment on an exchange between Dennett and Ruth Millikan. Responding to Dennett's claim that 'we can describe all processes of natural selection without appeal to . . . intentional language, but at enormous cost of cumbersomeness, lack of generality, and unwanted detail' (1990, p. 189), Millikan writes:

> if we dropped all talk of function or purpose in the biological world we would indeed be unable to discern most of the important patterns that are there. But . . . there is no need to drag the whole intentional stance into biology in order to perceive Nature's handiwork and the principles of natural design. (2000, p. 64)

I think that Millikan is partly right here. I agree with her, against Dennett, that there is no need to personify mother nature as an intentional agent in order to do evolutionary biology, nor to make sense of how biologists think and talk; indeed there is positive reason not to do this, as was stressed in our critique of agential thinking (type 2). But agential thinking (type 1) is a different matter. Treating an evolved organism as akin to an agent with a goal, and conceiving of its traits as means by which it tries to achieve its goal, does real work in biology; and where evolved behaviour is at issue, this fits naturally with making the organism the subject of intentional attributions and applying the organism-as-rational-agent heuristic. There *is* a pattern in nature that this helps us to discern, namely the organismic unity and integration that such attributions presuppose; and this is not captured by talk of Darwinian function alone.

1.9 Conclusion

We began by contrasting two possible attitudes towards agential thinking in biology: that it reflects our tendency to anthropmorphize, and that it is an apposite way of describing the objective biological facts. I suggest that the appropriate attitude depends on the type of agential thinking in question. Many organisms really do exhibit some agent-like attributes, namely goal-directedness and behavioural flexibility; and treating them as akin to agents trying to achieve an (ultimate) goal, in the context of evolutionary explanation, has a genuine point; for the unity-of-purpose that is thereby presupposed reflects an objective and important feature of evolved organisms. However, I see little reason to describe the *process* of natural selection in agential or intentional terms; this is more likely to mislead than enlighten.

2

Genes and Groups as Agents

2.1 Introduction

The previous chapter described a mode of thinking in evolutionary biology that we termed 'agential'. We distinguished two types of agential thinking: type 1, in which an evolved entity is treated as an agent with a goal and made the subject of agential/intentional attributions; and type 2, in which the evolutionary process itself is described in agential terms. We argued that agential thinking of type 2 can be insightful but is potentially misleading, while that of type 1 is a legitimate expression of adaptationist assumptions and plays a genuine intellectual role in biology. This chapter continues our exploration of agential thinking of type 1.

In typical biological applications of agential thinking (type 1), the agent is the individual organism and its goal is to maximize its lifetime reproductive success (or some other appropriate fitness measure). This reflects the fact that adaptationist explanations in biology are generally sought at the individual level. However, there are other candidates for the role of agent too. In the work of Dawkins (1976) and Haig (2012), agential thinking is applied to genes (or alleles); while in the literature on superorganisms and on collective decision-making in insect colonies, whole groups are treated as akin to agents with goals. The aim of this chapter is to explore agential thinking as applied to genes and groups, rather than individuals.

This raises issues closely related to the 'levels of selection' discussion in biology, that I have analysed previously (Okasha 2006). That discussion is about at which level(s) of the biological hierarchy the process of natural selection occurs, and which biological entities exhibit adaptations. Indeed an obvious suggestion is that once we have identified the relevant level of selection/adaptation in any particular case, this will immediately yield the right candidate for the role of agent, if we wish to apply agential thinking. I think this is partially correct; however, the matter is complicated by ongoing disagreement over how to identify levels of selection/adaptation, and indeed over what these terms mean. So I prefer to approach the matter differently, by direct consideration of biological work in which genes and groups are treated as agents, by invoking the unity-of-purpose constraint on agency from the previous chapter, and in the case of groups, by drawing on some ideas from the philosophy of social science.

The structure of this chapter is as follows. Section 2.2 considers genes as agents, with particular reference to Dawkins' arguments and to the phenomenon of intra-genomic conflict. Section 2.3 considers agential thinking as applied to whole groups and colonies, examines the unity-of-purpose constraint at the group level, and asks whether the choice of agent is an empirical or conventional matter. Section 2.4 explores a parallel with the discussion of group agency in social science, with reference to the doctrine of methodological individualism. Section 2.5 examines the biological veil-of-ignorance, which is one way that individuals' interests can be aligned, and explores a parallel with the veil-of-ignorance concept in political philosophy and economics. Section 2.6 concludes.

2.2 Genes as Agents

The genes in an organism sometimes 'disagree' over what should happen . . . some genes may want (or act as if they want) a male to produce lots of healthy sperm, but other genes in the same male want half the sperm to be defective. Some genes in a female want her to nourish all her embryos; others want her to abort half of them. (Burt and Trivers 2006, p. 1)

This quotation is from the opening paragraph of the book *Genes in Conflict*. The book contains numerous examples of agential thinking as applied to genes, which the authors use to make sense of a range of biological phenomena associated with intra-genomic conflict. These include sperm-killing, meiotic drive, cytoplasmic male sterility, sex-ratio distortions and more. These phenomena, which arise from conflict between the genes within the genome of a single individual, cannot be explained as ordinary individual-level adaptations, for they are typically harmful to the individual organism. By treating them as strategies designed by genes to further their own interests, Burt and Trivers are able to bring the phenomena within the Darwinian fold.

The *locus classicus* for the 'genes as agents' view is of course Dawkins' *The Selfish Gene* (1976). In his original case for 'genic selectionism', Dawkins argued that it offers a fully general way of understanding Darwinian evolution, but is especially well-suited to explaining two phenomena in particular: altruism and 'outlaw' genes. Altruism refers to actions that are costly for the actor but benefit others; it is found throughout the living world. Dawkins argued that altruism is anomalous from an organism's perspective but makes sense from the perspective of the gene that causes the altruism—it is trying to help copies of itself in other organisms. Outlaws are genes that evolve despite harming their host organism, by gaining a transmission advantage; they are exactly the genes that Burt and Trivers (2006) are describing in the quotation at the beginning of this section. Dawkins argued that outlaws, or 'selfish genetic elements' as they are today known, can *only* be understood from the gene's eye viewpoint; for they are often harmful at the level of the organism, group, or population.

To both phenomena, Dawkins applied agential thinking and the intentional idiom. He argued that the phenomena become intelligible if we put ourselves in the shoes of a gene and ask: 'How can I increase my representation in the gene-pool?' However, the two phenomena are quite different in kind (so should not be thought of as supporting a single doctrine of 'genic selectionism'). When a gene for altruism spreads, for example, by standard kin selection in which individuals preferentially assist their relatives, the selective mechanism is differential individual reproduction; that is, individuals bearing the altruistic gene leave more offspring than non-bearers, on average, in the global population. When an outlaw gene spreads, for example, a meiotic driver which biases segregation in its favour, no differential individual reproduction need be involved; and indeed outlaws often harm their host individual's reproductive output. Indicative of this difference is that outlaw genes create potential intra-genomic conflict, while genes for altruism typically do not (Ridley and Grafen 1981, Okasha 2002).

This distinction is sometimes marked by distinguishing 'selfish genes', which are ubiquitous, from 'ultra-selfish genes' (Dawkins' outlaws) which are relatively rare. Alternatively, we can distinguish gene-level selection, which refers to a particular type of selection process—selection between the genes within a single organism's genome—from the gene's eye perspective, which is a vantage point from which any selection process, at whatever level, can be viewed (Okasha 2006). When a trait spreads by ordinary individual-level selection, or by kin or group selection, we can still adopt the gene's eye viewpoint if we wish, since the net result is the spread of one genetic variant at the expense of its alleles; but only outlaws spread by the process of gene-level selection.

To see how this distinction relates to agential thinking, note that every gene is necessarily playing a zero-sum game against other alleles at the same locus: it can only spread in the population if they decline.[1] Therefore every gene, including outlaws, can be thought of as 'trying' to outcompete its alleles, or having this as its *ultimate* goal. This is the sense in which Dawkins can describe all genes as selfish. Genes can pursue their ultimate goal in different ways, that is, by pursuing a variety of *intermediate* goals. Most genes' intermediate goal is to enhance their host organism's fitness, which they achieve via their effect on the organism's phenotype. Other genes have as their intermediate goal to enhance the fitness of their host organism's relatives, by causing altruistic behaviour. However, a small handful of genes have a different intermediate goal, namely to increase their own transmission in their host organism's gametes, for example, by biasing segregation in their favour, or distorting the sex-ratio, or transposing to new sites in the genome. These are outlaws, or selfish genetic elements.

If outlaws are absent or are effectively suppressed, then the genes within a single organism have a common (intermediate) goal, so will cooperate: each gene can only

[1] Strictly the alleles at a locus play a unit-sum game, since their frequencies sum to one. Here I follow the standard game-theoretic convention of equating constant-sum and zero-sum games, since they are strategically equivalent (see Luce and Raiffa 1957, p. 158).

benefit itself by benefiting the whole organism. Agential thinking can then be applied to the organism itself. The organism's goal—maximizing its fitness—then equates to the intermediate goal of each of the genes within it. Each gene 'tries' to help its host organism (or its relatives) to survive and reproduce, as by doing this it can further its ultimate goal, namely outcompeting its alleles. In such cases, the phenotypic trait that a gene codes for can equally be thought of as an instrument by which the organism tries to achieve its goal, or by which the gene tries to achieve its intermediate goal. This is the sense in which Dawkins can describe the genic and the organismic perspectives as equivalent.

Outlaw genes typically harm their host organism; this induces selection for modifier genes at other (unlinked) loci for suppression of the outlaw, that is, intragenomic conflict. Such suppression often succeeds, which is why individual organisms are usually fairly cohesive entities that are not riven by internal conflicts, and can be treated as agents. If there is unsuppressed conflict, then different genes within an organism pursue intermediate goals that are mutually antagonistic—as in the examples of cytoplasmic male sterility in plants and sperm-killing in fruit-flies in section 1.6—so the organism displays traits that do not conduce to a single goal. The organism thus lacks the unity-of-purpose that is characteristic of agency. Agential thinking can still be applied, but only by treating the genes within the organism as the agents.

The moral, therefore, is that in applying agential thinking to genes we must attend to two distinctions. First, to say that genes 'want' to further their own replication, or encode strategies that further that goal, is ambiguous. Is this the logical point that the spread of any gene is necessarily at the expense of its alleles, or the empirical point that some genes spread via mechanisms that harm their host organism? Only in the latter case are we compelled to treat the gene rather than the organism as the agent, in order to apply agential thinking. Secondly, we need to distinguish ultimate from intermediate goals. In the spirit of Dawkins (1976), we may treat any gene as having the ultimate goal of spreading in a population. A gene pursues its ultimate goal via an intermediate goal, which is usually, but not always, to enhance its host organism's fitness. Where the genes within an organism share an intermediate goal, we may identify this with the whole organism's goal, and apply agential thinking to the organism itself.

The real merit of the 'genes as agents' concept, as I see it, is that it allows us to achieve an adaptationist understanding of a class of phenotypic traits that would otherwise seem puzzling. The traits in question are the phenotypic consequences of selfish genetic elements, or outlaws. These traits are detrimental to the fitness of the organisms that exhibit them, so cannot be explained as individual adaptations. By treating genes as agents pursuing their own interests, which need not be aligned with the interests of their host organism, we are able to bring these traits within the adaptationist fold, by explaining them as strategies that benefit the genes that cause them, rather than the organisms that exhibit them.

How does this case for treating genes as agents connect with the case for treating organisms as agents, discussed in section 1.5? Recall that we identified three distinct rationales in biology for treating organisms as agent-like: goal-directedness, behavioural flexibility, and adaptedness. We argued that the three rationales were independent though sometimes dovetail, and that the third was the relevant one in an evolutionary context. The point was that in a typical adaptationist explanation, a trait is explained by showing how it benefits the organism or furthers its evolutionary goal; thus the notions of agent, benefit, and goal are presupposed by such explanations. In the case of genes, we have argued that selfish genetic elements constitute the strongest argument for agential thinking. This is the analogue of the third rationale—adaptedness. The point is that selfish genetic elements have phenotypic effects which can be regarded as adaptations, but only if we apply the notions of agent, benefit, and goal to genes themselves.

What about the first and second rationales? Do these have analogues for genes? I think the answer is no. Genes themselves, as relatively inert molecules, do not exhibit goal-directed activity or behavioural flexibility as whole organisms do. (Though of course they *give rise* to such activity, via their causal influence on the organism's phenotype.) Indicative of this is that in the synchronic study of how genes work, in molecular biology, the notion of genes as agents has no currency and agential idioms are not found.[2] It is only in an evolutionary context that it makes sense to treat genes as agent-like and to credit them with goals and interests. By contrast, we argued that goal-directedness and behavioural flexibility would motivate treating organisms as agent-like even if creationism turned out to be true; these rationales are non-historical, deriving from observed properties of contemporary organisms.

I do not think that this disanalogy invalidates the case for treating genes as agents, in the particular context identified above as motivating this idea, namely supplying adaptationist explanations for the phenotypic traits to which selfish genetic elements give rise. For we have seen that the three rationales for the organism-as-agent concept are logically independent. That the first two do not apply to genes is thus no argument against applying agential thinking to genes in those circumstances when the third one does apply.

2.2.1 Ultimate versus intermediate goals

The idea of genes as agents, and the distinction between ultimate and intermediate goals, can be clarified using the rational choice formalism of section 1.7. As we saw, this formalism offers a quite general way of capturing agential idioms—the language of goals, interests, and preferences—as they are used in biology. For accuracy, here we talk about the goals of *alleles*, that is, particular genetic variants, rather than genes; this is what Dawkins (1976) was really talking about, as has long been recognized.

[2] Indeed we saw in section 1.2 that it is a moot point whether genes count as agents in even the minimal sense of an entity which *does* something.

Suppose there are n genetic loci in a population. Let m_i denote the number of alleles at the ith locus. The genetic composition of the population, at any time, may be described by a vector which specifies the frequency of every allele at every locus:

$$\mathbf{p} = \langle p_{11}, \ldots, p_{1j}, \ldots, p_{1m_1}, \ldots, p_{i1}, \ldots, p_{ij}, \ldots, p_{im_i}, \ldots, p_{n1}, \ldots, p_{nj}, \ldots, p_{nm_n} \rangle$$

where p_{ij} is the frequency of the jth allele at the ith locus. Thus $0 \leq p_{ij} \leq 1$ and $\sum_{j=1}^{m_i} p_{ij} = 1$ for all i.

Let X be the set of all possible genetic compositions of the population. Each allele has a preference over X, preferring vectors in which it has a higher frequency. Let P_{ij} and I_{ij} denote the strict preference and indifference of the jth allele at the ith locus, respectively. An allele's ultimate goal may then be captured as follows. For all vectors $\mathbf{q}, \mathbf{r} \in X$:

$$\mathbf{q}P_{ij}\mathbf{r} \text{ if and only if } q_{ij} > r_{ij}$$
$$\mathbf{q}I_{ij}\mathbf{r} \text{ if and only if } q_{ij} = r_{ij}$$

where q_{ij} and r_{ij} denote the frequency of the jth allele at the ith locus in the vectors $\mathbf{q}$ and $\mathbf{r}$, respectively.

Two consequences are immediate. First, no two alleles, at the same or different loci, have identical preferences over the entire set X—each allele 'cares about itself'. Secondly, at any locus i, if there are exactly two alleles then there is a pure conflict of interest, that is, $\mathbf{q}P_{i1}\mathbf{r}$ if and only if $\mathbf{r}P_{i2}\mathbf{q}$. If there are more than two alleles at the ith locus, then for any vectors $\mathbf{q}, \mathbf{r} \in X$ such that $q_{ix} = r_{ix}$ for all $x \neq 1, 2$: $\mathbf{q}P_{i1}\mathbf{r}$ if and only if $\mathbf{r}P_{i2}\mathbf{q}$. These two consequences capture the meaning of the claim that all alleles are 'selfish' in Dawkins' sense.

To capture an allele's intermediate goal, consider a focal organism. The organism's lifetime reproductive output r is the number of successful haploid gametes it produces, which ranges from 0 to a maximum of c. Let s_{ij} be the frequency of the jth allele at the ith locus found within the organism (= 0, 0.5 or 1 for diploids); let t_{ij} be its frequency among the successful gametes; so the total number of copies of the allele in the organism's successful gametes is $n_{ij} = r \cdot t_{ij}$.

The organism's gametic output is summarized by the vector:

$$\mathbf{n} = \langle n_{11}, \ldots, n_{1j}, \ldots, n_{1m_1}, \ldots, n_{i1}, \ldots, n_{ij}, \ldots, n_{im_i}, \ldots, n_{n1}, \ldots, n_{nj}, \ldots, n_{nm_n} \rangle$$

where $\sum_j n_{ij}$ is constant for all i, and equals r. Note that $\mathbf{n} = r\mathbf{t}$, where $\mathbf{t}$ is the corresponding vector of allele frequencies, describing the allelic composition of the successful gametes.

Let Y be the set of all possible gametic outputs (for variable r). The intermediate goal of any allele within the organism is to leave as many copies as possible in the organism's successful gametes. Therefore, the preferences of the ith allele at the jth locus within the organism must satisfy:

$$\mathbf{a}P_{ij}\mathbf{b} \text{ if and only if } a_{ij} > b_{ij}$$
$$\mathbf{a}I_{ij}\mathbf{b} \text{ if and only if } a_{ij} = b_{ij}$$

for all vectors $\mathbf{a}, \mathbf{b} \in Y$. Note that an allele's intermediate goal is described by a preference over the possible gametic outputs of its host organism Y, while its ultimate goal is described by a preference over the set of population compositions X.

Focusing on the intermediate goal, any two alleles within the organism, at the same or different loci, have mixed interests, that is, their preferences co-incide over some pairs of gametic outputs but disagree over other pairs. (This is because the number of successful gametes may vary between elements of Y). Thus there is potential for within-organism conflict; however, it may be suppressed by limiting the alleles' ability to influence which member of Y obtains.

To capture this point, suppose that meiosis is constrained to be fair. This means that Y is no longer the relevant choice set. For each vector in Y specifies the number of successful gametes r and their allelic composition; but given fair meiosis, any allele in an organism can influence which member of Y obtains only by influencing r. The relevant choice set is thus simply the set of possible successful gamete numbers $\{0, \ldots, c\}$, and over this set all alleles have a common interest, preferring more to less.

Finally, suppose that the organism's own 'goal' is to maximize r—its number of successful gametes—as is reasonable in many evolutionary scenarios. Then, the link between the organism's goal and the goals of its constituent alleles is immediate, if the choice set is restricted in the above way. The alleles have a common intermediate goal that coincides with their organism's goal.

The purpose of this formalism is not to derive new consequences, but rather to make precise what we knew already. The formalism allows us to explicitly distinguish genes' ultimate goals from their intermediate ones, in terms of preferences over different sets, and captures one way in which alignment of the interests of the genes within an organism is achieved, namely by restricting their ability to influence the allelic composition of the organism's successful gametes.

2.2.2 *A metaphor without a cause?*

The idea of genes as agents with strategies has often been criticized, by philosophers and others. These criticisms are a mixed bag. Many have been satisfactorily rebutted, for example, Midgley's insistence that Dawkins' attribution of 'selfishness' to genes makes no sense, even as metaphor; and the accusation that this way of thinking leads to genetic determinism (Midgley 1979, Dawkins 1981). A more serious objection has recently been developed by Godfrey-Smith (2009). While acknowledging the heuristic utility of the gene's eye view in certain cases, Godfrey-Smith regards the genes-as-agents concept as ultimately harmful, and a source of scientific error.

Godfrey-Smith is partly objecting to agential thinking in biology *tout court*, of which he takes genes-as-agents to be the paradigm case. He regards 'the language of agency and benefit', applied to evolved entities of any sort, as a short-hand that owes more to what we humans find psychologically potent than to its particular aptness for describing evolution (2009, p. 164). Agents pursuing goals are inessential to Darwinian theory, Godfrey-Smith argues, evidenced by the fact that there is a 'style of selectionist thinking' in which agents and agential metaphors play no role. This style is typified by questions of the form: 'suppose a population was like this, and such-and-such a mutation appeared, what would happen to it?' (2009, p. 145).

In a way this is true; however, what Godfrey-Smith here describes is a way of thinking about *selection*, not adaptation. Darwin's original argument, of course, was that selection leads to adaptation, or the appearance of design, in nature. Certainly one can describe the process of natural selection itself without invoking agents, goals, or interests, as for example in the language of population genetics; at root, it is about the spread of some genetic variants and the decline of others. Indeed in section 1.4, we argued that agential thinking as applied to the process of selection is misleading. However, this does not show that one can characterize Darwin's *explanandum*, the phenomenon of phenotypic adaptation, without invoking a notion of benefit and thus beneficiary, or agent. The 'fit' of organism to environment at which Darwin marvelled is about an organism having traits that benefit it, or further its biological goal; the agential idiom is hard to avoid here.

Godfrey-Smith has another objection which is specific to Dawkins' position. He argues that Dawkins' commitment to an agential picture of evolution leads him to an abstract characterization of Darwinian evolution which is overly strict and not universally applicable. Dawkins' view is that evolution by natural selection requires 'long-term persisting entities', or lineages of replicators, such as genes. Individual organisms and their genotypes are not enough, even if their traits are heritable, for they are transitory entities lasting a mere generation, and thus incapable of playing the role of long-term beneficiary of the evolutionary process. As Godfrey-Smith sees it, Dawkins has an a priori commitment to there being such a long-term beneficiary, which derives from agential thinking, and which leads him to think that replicators are an essential ingredient of all Darwinian evolution. This conclusion is clearly mistaken according to Godfrey-Smith, since some *bona fide* Darwinian processes, such as cultural evolution, do not involve replicators.

Godfrey-Smith has a good point here; however, the aspect of Dawkins' views at which it is directed is not the aspect that I would emphasize. The real merit of the 'genes as agents' idea, as I see it, is that it enables us to make sense of intra-genomic conflict and its phenotypic consequences, and thus to extend the adaptationist paradigm to a range of phenomena that would otherwise be baffling. Sperm-killing by other sperm, sex-ratio distortion, transposable elements, and cytoplasmic male sterility are all features that make sense once we see them as strategies used by genes to benefit themselves, rather than their host organisms. These features do yield to adaptationist explanation, and the characteristic type of understanding that comes in its wake, but only when genes are treated as the beneficiaries. This ties in with Gardner and Welch's observation that the early scientific attempts to understand selfish genetic elements, prior to Dawkins' gene's eye view, led to 'conceptual tangles' (2011, p. 180).[3]

Intra-genomic conflict formed part of Dawkins' case for the gene's eye view, as we have seen; but he also emphasized unrelated phenomena such as kin-directed altruism, as well as the points about high-fidelity replication and long-term persistence that Godfrey-Smith criticizes. However, these ideas are logically separable. It is only the first which constitutes a compelling case for treating genes as agents with strategies, and making them the subject of agential/intentional attributions. I agree that there is no a priori reason for conceiving of Darwinian evolution this way. Rather, it is the empirical fact that organisms exhibit certain features whose adaptive significance can only be understood by treating them as for the benefit of the genes that cause them, rather than for the organisms' benefit, that legitimizes treating genes as agents.

This conclusion is an instance of a general moral that was hinted at previously and is developed in detail in chapters 3 and 4. The moral is this: to the extent that agential thinking in evolutionary biology is justified, it is empirical facts about the world that justify it, and not some overarching theoretical principle about how natural selection works.

2.3 Groups as Agents

The single best demonstration of the superorganismic nature of a honeybee colony is the ability of a honeybee swarm to function as an intelligent decision-making unit when choosing its new home.

(Seeley 2010)

In contemporary biology the idea of groups as agents arises in two distinct though related bodies of literature. The first concerns collective decision-making, in which a group of individuals share information and engage in coordinated group activity as a result; this activity may relate to foraging, movement, or defence, for example.[4] Collective decision-making occurs in diverse taxa including mammals, birds, and microbes, but has been best documented in social insects, particularly ants and bees. For example, Seeley (2010) describes the remarkable way in which honey-bee colonies choose a new nest site, by using a 'democratic decision-making process' (p. 1). This process involves evaluation of competing sites, exchange of information, discussion of the options, consensus building, and coordinated navigation to the chosen site. Given such impressive coordination it is natural, indeed almost inevitable, to treat the whole colony as akin to a single agent with a goal. We can sensibly ask what the colony is *doing*, what it is *trying* to achieve, and what it *knows* about the quality of the various nest sites. Thus Seeley makes free use of the language of decision, choice, and preference—paradigmatic agential idioms—in relation to whole colonies.

[3] Gardner and Welch (2011) cite Gershenson (1928), pp. 506–7, as an example here.

[4] See Ross-Gillespie and Kummerli (2014) for a useful overview.

The second body of literature is in social evolution theory, and concerns group-level adaptations: features that evolved because they benefit groups. How prevalent such adaptations are in nature, and indeed whether they exist at all, is an old debate (Williams 1966). But there are compelling examples, paradigmatically in eusocial insect colonies. Such colonies often display complex adaptive organization, and individual insects exhibit features, morphological and behavioural, which appear designed to benefit the whole colony. Worker sterility is the obvious example. Sterile workers are unable to have (or prevented from having) offspring of their own, so devote their whole lives to assisting the colony, by feeding the larvae, foraging, and protecting the nest. Such phenomena motivate the idea that eusocial insect colonies are 'superorganisms' rather than societies. This idea remains controversial, thanks to the over-enthusiasm of its original advocates, but most biologists would accept that it has approximate validity in certain cases—in particular where internal conflict is effectively suppressed, and the colony functions as an integrated whole.[5] In such cases, the notions of benefit, goal, and interest apply at the group level.

These two rationales for regarding groups as agent-like are related though distinct. They are related since the ability of a group to make collective decisions is often likely to be a group-level adaptation. But they are distinct since the former derives from non-historical attributes of certain groups, that is, their exhibiting behaviour which invites explanation in agential/intentional terms; while the latter is explicitly evolutionary, deriving from the existence of features which evolved because they confer a benefit at the group level. In effect, the former is the group-level analogue of goal-directedness and behavioural flexibility—our first two rationales for treating organisms as agents; while the latter is the group-level analogue of adaptedness—our third rationale for the organism-as-agent concept. As discussed previously, these rationales are independent, though sometimes dovetail. Our focus here will be on the third, evolutionary rationale for applying agential thinking to groups.

Recall the unity-of-purpose constraint on attributions of agency described in section 1.6. For an entity to be be treated as an agent, for the purposes of adaptationist theorizing, its evolved traits must contribute to a single overall goal, and thus have complementary rather than antagonistic functions. We saw how within-organism conflict, unless effectively suppressed, can threaten the status of individual organisms as agents; a similar moral applies to groups. With a few exceptions, for example, certain insect colonies and marine invertebrate clones, most groups in nature seem unlikely to satisfy the constraint, as within-group selection, and thus conflict, are common. This suggests that the scope for applying agential thinking to groups is more limited than for individual organisms, a conclusion borne out by biological practice.

[5] The superorganism concept originated in the work of the ecologists Allee (1931) and Emerson (1939). For more recent discussions, see Wilson and Sober (1989), Seeley (1996), Gardner and Grafen (2009), and Hölldobler and Wilson (2008).

This is not to say that group or multi-level selection is rare, but only that it does not usually lead groups to exhibit the degree of internal harmony that a typical individual has. Indeed in a sense this is a definitional rather than a substantive truth, since where groups *do* evolve a high degree of cooperation and functional integration, we tend to elevate them to the status of 'individuals' and regard their members as parts of a single whole (Queller 1997). Thus a multi-celled organism, the paradigm biological 'individual', began as a loose aggregate of free-living cells, gradually evolving into a cohesive unit; and the eukaryotic cell itself evolved from the union of free-living prokaryotes. This is a familiar point in the literature on evolutionary transitions—which studies the gradual process by which larger biological units evolved from aggregations of smaller units (Buss 1988, Maynard Smith and Szathmáry 1995, Bourke 2011). The point teaches us that wholesale dismissal of the idea of selection for group advantage cannot be correct.

I have analysed the evolutionary transitions question previously, as have many other philosophers.[6] Rather than rehearse this discussion, here I want to focus on one issue specific to agential thinking, namely the idea that the choice of 'agent' may sometimes be a matter of convention, rather than objective fact.

2.3.1 Conventionalism about agents?

In the levels-of-selection debate, it is often suggested that the choice between alternative levels is a matter of convention, not empirical fact, at least in some cases. In particular, the idea that the evolution of social behaviour can equally be explained in terms of multi-level selection—which invokes selection at individual and group levels—or in terms of kin selection/inclusive fitness—which invokes selection only at the individual level—has achieved orthodoxy in recent discussions. Proponents of this view describe kin and multi-level selection as 'equivalent'; the choice between them is a matter of pragmatic convenience, not empirical fact, they argue.[7] It is natural to ask whether something similar applies to agential thinking. Might the choice between treating individuals and groups as agents, in the context of adaptationist theorizing, also be conventional?

A putative argument for this form of conventionalism goes as follows. Consider an altruistic trait such as the honey bee's barbed stinger. Since a bee dies when it uses its sting to repel an aggressor but the colony as a whole benefits, the stinger can be regarded as a colony-level adaptation: a means by which the colony protects itself, thus increasing colony fitness. However, Hamilton's 'inclusive fitness' concept shows a way to re-interpret the stinger as an individual-level adaptation. As Hamilton (1964) taught us, an individual can transmit its genes to the next generation indirectly, via helping its relatives leave offspring. So even though the honey bee harms itself by

[6] See Okasha (2005, 2006), Godfrey-Smith (2009), Calcott and Sterelny (2011), Clarke (2014), and Birch (2017).

[7] This equivalence thesis is found in Grafen (1984), Queller (1992), Lehmann *et al.* (2007), West *et al.* (2007), Marshall (2011), Frank (1998, 2013), and others. See Okasha (2015) for extended discussion.

stinging, thus reducing its personal fitness, it may enhance its inclusive fitness—a measure which takes account of an individual's contribution to its relatives' reproductive success. Therefore, the argument goes, it is a matter of convention whether the honey bee's stinger is an individual or a colony-level adaptation; and thus it is similarly conventional whether the individual bee or the colony is treated as the agent with a goal.

Before assessing this argument, two points deserve note. Firstly, the Hamiltonian strategy for restoring individuals as agents—taking their goal to be inclusive rather than personal fitness—is interestingly analogous to a strategy used by certain social scientists (e.g. Fehr and Schmidt 1999, Bolton and Ockenfels 2000). Faced with apparently irrational human behaviour, for example, 'altruistic punishment' in which people pay money to punish those who have violated social norms, these theorists argue that the behaviour is actually rational, that is, utility-maximizing, relative to a different choice of utility function. For example, perhaps agents care about equity and reciprocity as well as the size of their personal payoff? So by positing a different goal, observed behaviour may be reconciled with the assumption of individual rationality, which was exactly Hamilton's argument too. This parallel is explored further in section 7.2. Secondly, note the interaction between choice of agent and choice of goal, in applying agential thinking. In order to determine whether a given trait, such as the honey bee's stinger, can be validly understood in terms of individual agency, it is essential to consider a suitable choice of individual goal. This moral applies to agential thinking quite generally.

The conventionalist claim says that while Hamilton's strategy for interpreting altruistic traits as individual adaptations is valid, an equally valid alternative is to regard such traits as group adaptations, designed to benefit the whole colony. To probe this issue further, we need to consider in more detail the suggestion that kin and multi-level selection are equivalent. For it is this 'equivalence thesis', as I will call it, that underpins the conventionalist argument about agency sketched above.

2.3.2 *The equivalence thesis*

Consider a structured population of the sort depicted in Figure 2.1, in which individuals live in groups within which social interactions occur. A population structure of this sort characterizes many social species, and often leads to non-random associations between individuals, which is the key requirement for altruistic traits to evolve. Now the equivalence thesis stems from the fact that in such a population, a correct expression for gene frequency change may be written in either multi-level or kin selection terms. The former attaches a fitness value to both groups and individuals, the latter only to individuals.

To make this concrete, assume for simplicity that individuals are haploid and that mutation is absent; so individuals faithfully transmit their genotype to their offspring. Consider a gene that codes for a social trait, such as an altruistic behaviour. Let p_i denote the genetic value of the ith individual in the global population, where

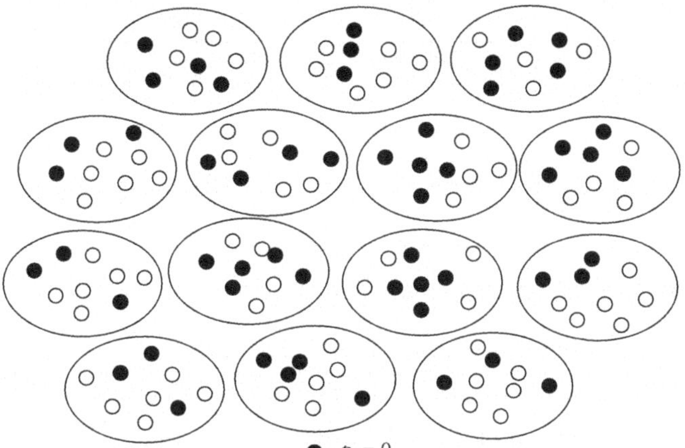

Figure 2.1. Individuals in a group-structured population

$p_i = 1$ if the individual has the gene, and $p_i = 0$ otherwise; so the average genetic value, denoted p, is simply the population-wide frequency of the gene. Let w_i denote the realized fitness of the *i*th individual, that is, its number of surviving offspring, and let $\overline{w}$ denote the average fitness. The change in gene frequency over a generation can then be written using a simplified form of the Price equation:

$$\Delta p = \frac{\mathrm{Cov}(w_i, p_i)}{\overline{w}} \tag{2.1}$$

where the covariance is taken over all individuals in the population (Price 1970). Equation (2.1) tells us that whether the gene for the social trait increases or declines over a single generation depends on whether there is a positive or negative association between the social trait and fitness in the global population. This is highly intuitive, capturing the core neo-Darwinian idea that genes associated with higher individual fitness will spread by natural selection.

In effect, multi-level and kin selection represent alternative ways of partitioning the total change Δp into two components. The former partitions Δp into within-group and between-group components, capturing the idea that selection acts at two hierarchical levels. The latter partitions Δp into personal and social (or direct and indirect) components, capturing the idea that an individual's fitness depends on both its own traits and its social partners. This yields a version of Hamilton's rule, the famous $rb > c$ criterion for the spread of an altruistic trait, in which c measures the cost of the trait for the individual that expresses it, b the benefit of the trait for the individual's social partners, and r the genetic correlation (or relatedness) between social partners (Hamilton 1964). The two partitions are written below; the multi-level partition is explained in Box 2.1, while the kin selection partition is explained in chapter 5, section 5.3.

$$\overline{w}\Delta p = \overbrace{\text{Cov}(W_k, P_k)}^{\text{between-group}} + \overbrace{E[\text{Cov}(w_{jk}, p_{jk})]\}}^{\text{within-group}} \tag{2.2}$$

$$= \overbrace{(-c)\text{Var}(p)}^{\text{personal}} + \overbrace{rb\text{Var}(p)}^{\text{social}} \tag{2.3}$$

Box 2.1. Multi-level selection partition

$$n_k = \text{number of individuals in } k\text{th group}$$
$$p_{jk} = \text{genetic value of } j\text{th individual in } k\text{th group}$$
$$w_{jk} = \text{fitness of } j\text{th individual in } k\text{th group}$$
$$P_k = \frac{1}{n_k}\sum_j p_{jk} = \text{average genetic value of } k\text{th group}$$
$$W_k = \frac{1}{n_k}\sum_j w_{jk} = \text{average fitness of } k\text{th group}$$

The overall covariance $\text{Cov}(w_i, p_i)$ in equation (2.1) can be written:

$$\text{Cov}(w_i, p_i) = \text{Cov}(W_k, P_k) + \text{E}[\text{Cov}(w_{jk}, p_{jk})] \tag{2.4}$$

Here $\text{Cov}(W_k, P_k)$ is the covariance between the group means, and $\text{E}[\text{Cov}(w_{jk}, p_{jk})]$ is the average across groups of the within-group covariances; both terms are weighted by group size. Substituting equation (2.4) into equation (2.1) yields equation (2.2) in the text.

See Price (1972a) for original derivation of equation (2.4), and Okasha (2006) for discussion.

Proponents of the equivalence thesis then argue that since equations (2.2) and (2.3) are both mathematically correct, there is no fact of the matter about whether kin selection or multi-level selection is at work; these are equivalent descriptions of a single selective process.[8] The core kin selection idea—that a social trait will spread if the inclusive fitness effect ($rb - c$) is positive—and the core multi-level selection idea—that a social trait will spread if between-group selection is stronger than within-group—are thus different ways of the saying the same thing, it is claimed.

Is this a good argument for conventionalism about the choice of agents? There are a number of tricky issues here. Firstly, note that although there is undeniably a formal

[8] This way of arguing for the equivalence of kin and multi-level selection originated with Hamilton (1975), and is often appealed to in the social evolution literature, for example, Queller (1992), Frank (1998), and Marshall (2015).

equivalence between multi-level and kin selection, this does not show that there is a causal-explanatory equivalence. Certainly, Δp can be decomposed using either equation (2.2) or (2.3); but these equations are silent about causality. To conclude that if the social trait spreads, that is, if $\Delta p > 0$, there is no fact about whether this is *because* it enhances group fitness, or *because* it enhances the inclusive fitness of individuals, is not necessarily warranted, as it presumes that both partitions are equally true to the causal facts. But this need not be so. It is quite conceivable that in certain biological cases the multi-level selection partition (2.2) is a better representation of the causal facts, while in other cases the reverse is true. (For example, if the groups in the population have been arbitrarily defined but lack any 'biological reality', then the multi-level partition will arguably mislead as to the true causal structure of the selection process (Okasha 2015).) The key point is that selection is a causal notion; so the equivalence, or otherwise, of two selectionist explanations cannot be a purely formal matter.

Secondly, even leaving this point aside, note that equations (2.2) or (2.3) both describe the outcome of selection over a single generation, not adaptation. It may seem tempting to infer from equation (2.3), the kin selection partition, that social evolution will lead individuals to evolve traits that maximize their inclusive fitness, and from equation (2.2) that social evolution will lead groups to evolve traits that maximize group fitness, but neither inference is in fact valid without further assumptions. One cannot directly deduce a claim about maximization, or the 'goal' towards which an evolved trait will appear designed, from an equation giving the change in frequency of the trait over a single generation. (Essentially, this is because the right-hand side terms may be frequency-dependent; so the one-generation change may not determine the full evolutionary dynamics.) Indeed the idea that individuals' maximizing their inclusive fitness is the predicted outcome of natural selection on social traits, though often asserted, is a source of ongoing controversy in the literature. Critics have argued that an individual's inclusive fitness is only a well-defined quantity in certain simple cases, and that the maximization claim is similarly limited in scope.[9] This issue is examined in chapter 5.

Thirdly and most importantly for now, even if we grant that a particular trait, such as a honey bee's stinger, may equally be viewed as an individual or a group adaptation, it is a further step to the conclusion that individual and group are equally good candidates for the role of agent. For as stressed in section 1.6, agential thinking applies to whole entities—individuals or groups—not to their traits, and requires that all traits conduce to a common goal. Now while honey-bee colonies are remarkably cohesive entities, most social groups are less so. Rather, they exhibit some traits that are group-beneficial but others that are not, for example, the conflicts in ant colonies between queens and workers over the sex-ratio (Ratnieks *et al.* 2006). The same is

[9] See for example Nowak *et al.* (2010), Allen and Nowak (2015), and Okasha and Martens (2016a).

true of most vertebrate social groups. If we focus on a single pro-social trait, it may be possible to explain it in terms of either individual or group advantage, but the same need not be true of other traits. This tells against applying agential thinking to the group, as the required unity-of-purpose will often be lacking.

To illustrate this point, consider colonies of the stingless bees *Melipona*. Sophisticated colony defence mechanisms exist in the *Melipona*, involving both physical and chemical aggression, and bees often die defending their colony (Shackleton *et al.* 2015). Thus colony defence appears to confer an adaptive benefit on the whole colony, so is a plausible candidate for being a group-level adaptation. But now consider the fact that up to 25% of the larvae in a *Melipona* colony may develop into queens rather than workers. This is highly wasteful, as it is far more than the number needed for colony reproduction by swarming, and the excess queens are often killed by other workers (Ratnieks *et al.* 2006). So in some respects the colony resembles an agent with a single objective, but in others it resembles a collection of agents with conflicting objectives.[10] Care must therefore be exercised in applying agential thinking to the whole colony, lest we fallaciously infer that producing an excess of queens must somehow promote the colony's goal.

This conclusion—that many biological groups lack the unity-of-purpose necessary for agency—tallies with a conclusion reached by Gardner and Grafen (2009) in an influential paper. They argue that only clonal groups, and groups in which reproductive competition is fully suppressed, qualify as agent-like. (Suppression of reproductive competition means that all individuals within a group have identical expected reproductive success, so there is no within-group selection and thus no conflict.) However, while our conclusion was arrived at by means of a philosophical point—internal conflict undermines the unity-of-purpose that agential thinking presumes—combined with an empirical observation—such conflicts are found in many biological groups—Gardner and Grafen take a different approach. Using Grafen's 'maximizing agent' concept, they offer a *proof* that only clonal or fully reproductively suppressed groups pass muster.[11] I now turn briefly to their argument.

2.3.3 Gardner and Grafen on group agents

Gardner and Grafen's argument works by frameshifting an argument originally developed by Grafen for individuals up to the group level. Grafen tries to justify the biological practice of treating individual organisms as if they were agents trying to maximize their fitness (Grafen 2002, 2006, 2014a). The logic of Grafen's argument is explored in detail in section 4.2. For the moment, the key point is that Grafen derives the general result that the 'individual-as-maximizing-agent' analogy, as he calls it, is valid as long as within-individual selection is absent, but not otherwise. This is quite intuitive, for such selection, for example in the form of meiotic drive

[10] Strassman and Queller (2007) refer to certain insect colonies as 'divided organisms'.
[11] Okasha and Paternotte (2012) offer a critical re-assessment of this proof.

or sex-ratio distortion, often leads to intra-genomic conflict, which as we have seen undermines the prerequisite for treating individuals as agents. In essence, Gardner and Grafen (2009) apply the same logic at the group level, concluding that groups qualify as 'maximizing agents' if and only if within-group selection is absent.

Under what circumstances will within-group selection be absent? In the context of the simple model above, the answer is obvious from inspection of equation (2.2), the multi-level selection partition. Logically there are just three possibilities. One is that the groups contain no genetic variance (i.e. $\mathrm{Var}(p_{jk}) = 0$ for all k), which corresponds to clonality. A second is that within-group fitnesses are equalized (i.e. $\mathrm{Var}(w_{jk}) = 0$ for all k), which corresponds to repression of competition within groups. A third is that within-group genetic variance and fitness variance both exist but are entirely uncorrelated; though empirically this is unlikely to be true for every gene in a population. Gardner and Grafen focus on the first and second possibilities, concluding that group adaptation and group agency are valid concepts when groups are clonal, or exhibit suppression of reproductive competition, but not otherwise.

This is an interesting analysis, and ties in with the empirical point that the most cohesive groups in nature are often those in which genetic relatedness is high, thus approximating clonality, or ones in which policing mechanisms prevent internal competition, thus suppressing fitness-variance. However, I think that Gardner and Grafen's conclusion is problematic, for it runs into a famous problem noted by G. C. Williams (1966) in his famous critique of the 'good of the group' tradition in evolutionary biology. It is instructive to spell this point out.

One of Williams' arguments was that there is a crucial difference between genuine group adaptation and what he called 'fortuitous group benefit'. His point was that even if a trait is beneficial for a group, it does not follow that it evolved for that reason; the group benefit may simply be a byproduct of the benefit that the trait confers on individuals. Thus a deer herd may benefit from the fleetness of its members, but this does not make it an adapted herd; rather it is a herd of adapted individuals. Williams argued, surely correctly, that a trait should be only be considered a group-level adaptation if it evolved *because* it benefits the group. Mere incidental benefit to the group does not count.

A corollary of Williams' point, long recognized through strangely overlooked in recent discussions, is that the multi-level Price equation (2.2) is not an infallible guide to the levels of selection, as it will tend to over-diagnose group selection.[12] That is, a positive covariance between a group trait and group fitness need not imply that the former causes the latter; it may be a side-effect of lower-level selection. An example of this is an individual trait which is entirely non-social, that is, only affects the fitness of individuals that bear it, but exhibits considerable between-group variance (for whatever reason). By Williams' lights such a trait is not a group adaptation, and

[12] This point was first made by Nunney (1985) and Heisler and Damuth (1987), and is emphasized in Okasha (2006).

does not evolve by group-level selection; even if there is a strong positive covariance between the trait's incidence in a group and the fitness of the group.

The same point causes trouble for Gardner and Grafen's analysis, in particular for their conclusion that clonal groups constitute the best-case scenario for group adaptation. Consider a clonal group that is entirely non-social, that is, its constituent individuals engage in no fitness-affecting interactions, so the reproductive prospects of any individual are unaffected by its group context. (Certain marine invertebrate clones, for example, bryozoans, are possible examples here.) Thus the traits that evolve will do so because of the benefit they confer on individuals; any group benefit is entirely fortuitous. This is exactly the sort of scenario in which talk of group adaptation is unhelpful (and in which the multi-level selection partition (2.2) is causally misleading); yet Gardner and Grafen's analysis yields the result that groups, not individuals, are the 'maximizing agents' in this example.

This untoward result arises, at root, from the fact that selection and adaptation are causal concepts, and in general correlational criteria can at best provide necessary conditions for causal relations to obtain, not sufficient ones. In particular, the notion of selection captured by the Price equation is trait-fitness covariance, or correlation; and as we will see in section 4.2, Grafen's analysis of the circumstances under which the 'maximizing agent' analogy applies rests essentially on the Price equation, so inherits this problem. For many purposes this does not matter, but in the context of structured populations and multi-level selection it assumes a particular importance, given the ever-present possibility that a covariance at the group level may be 'caused from below', that is, it may be a byproduct of selection on individuals.

I do not think that this invalidates Gardner and Grafen's analysis altogether; rather, it shows that their methodology—examining the conditions under which Grafen's 'maximizing agent' analogy obtains when groups are the agents—can at most identify necessary conditions for a group to count as a well-adapted unit, and thus for agential thinking to be applicable at the group level, but not sufficient conditions. It is quite plausible that for a biological group to usefully be treated as agent-like, for the purposes of adaptationist theorizing, within-group selection must be absent (or nearly enough); indeed this ties in precisely with our unity-of-purpose constraint on agency. It is much less plausible that *whenever* a biological group exhibits no within-group selection it can thereby be treated as agent-like; for this fails to respect Williams' distinction.

2.4 Group Agency in Social Science

The issue of individuals versus groups as agents, in the biological sense considered here, has an interesting parallel in the philosophy of social science. In that field there is an old but ongoing debate over whether social collectives—entities such as juries, committees, and corporations—can ever legitimately be treated as agents. In everyday parlance it is common to describe social collectives in agential terms and to make them the subject of intentional attributions. Thus we say that the jury believed the defendant, the committee decided to call a fresh enquiry, and the corporation wished to increase its market share. However, a venerable tradition holds that such talk is a mere *façon-de-parler* and should not be taken at face value. Only individuals have intentional states and perform actions according to this tradition, so social phenomena cannot be explained in terms of the intentions, goals, or actions of collectives. This is an essential part of the doctrine known as 'methodological individualism' in the philosophy of social science.

The debate over methodological individualism and the status of collective entities has many points of contact with the biological debate over individual versus group adaptation. In particular, opponents of group-level adaptationism in biology, such as G. C. Williams, use arguments that closely parallel those used by methodological individualists in social science, while defenders of group adaptationism argue similarly to those who defend the view that social collectives can be agents. This parallel is interesting for its own sake, and also connects with the discussion of adaptationist and psychological explanation from section 1.8. To explore the parallel, I start by briefly outlining the traditional case for individualism in social science.

Methodological individualism is a prescription about the *form* that social-scientific explanations should take. It says that all social phenomena should be explained in terms of the actions of the individuals that ultimately give rise to them. For example, suppose we wish to explain why US–Soviet relations thawed in the late 1980s. One might cite broad geopolitical factors, for example, the rise in foreign investment in the Soviet Union, or the shift from military to civilian expenditure. Methodological individualists would regard these explanations as inappropriate, or at least incomplete. A proper explanation should cite the actions of the particular individuals, for example, Gorbachev and Reagan, which ultimately led relations to improve. Only then can we attain a genuine understanding of why the social phenomenon occurred.

Methodological individualism derives from the work of the sociologist Max Weber. In *Economy and Society*, Weber (1922) wrote: 'collectivities must be treated as solely the resultants and modes of organization of the particular acts of individual persons, since these alone can be treated as agents in a course of subjectively understandable action' (p. 13). Later proponents of the doctrine included R. Hayek, R. von Mises, and K. Popper. The doctrine has had considerable influence on social scientific practice. Much of neo-classical economics adheres to individualist strictures, though in fields such as sociology and anthropology the picture is more mixed.

Importantly, methodological individualism does not simply stem from a commitment to reductionism, that is, explaining the properties of wholes in terms of their parts. Rather, Weber's demand that social phenomena be explained from an individualistic basis stemmed from a commitment to a particular mode of explanation, namely the intentional-psychological mode (Heath 2015). The idea was to show how social phenomena result from the deliberate, conscious actions of agents, as only this would furnish a true understanding of why they occurred. Since collective entities such as committees and firms are not agents with mental states and do not perform actions, they should not be appealed to in social-scientific explanation, Weber argued.

A different aspect of methodological individualism is opposition to 'functionalist' explanations in the Marxist tradition (Elster 1982, 1985). Marxist explanations often appeal to the 'interest' of social groups, for example, in the suggestion that it is in the interests of capitalists to keep wages low. This explanation is suspect: an individual capitalist might well profit from raising wages slightly to attract the best workers. The point is that the interests of one individual capitalist need not coincide with the interests of capitalists as a whole. By attempting explanations in terms of 'capitalists' and 'workers' en masse, Marxists were led to overlook collective action problems, Elster argued. Thus methodological individualism serves as an antidote to fallacious reasoning.

To see the biological parallel, consider again G. C. Williams' critique of the 'good of the group' tradition in evolutionary biology. Williams (1966) argued that true group adaptations, as opposed to fortuitous group benefits, are rare, and that group selection is an inherently weak evolutionary force. Williams' argument is structurally parallel to Weber's. Williams is concerned with a particular mode of explanation, namely the Darwinian or adaptationist mode. His claim is that only individuals exhibit adaptations and thus that Darwinian explanations should be individualistic: evolved traits should be explained by the advantage they confer on individuals, not groups. Similarly, Weber argues that only individuals are agents, thus social scientific explanations should be individualistic. The abstract logic of the argument—a commitment to a particular mode of explanation, plus a claim that the property required by this mode is only possessed by individuals—is identical in the two cases.

In fact the parallel is deeper than this, since the two modes of explanation are intimately related. We saw in section 1.8 that adaptationist explanation is rationalizing: it shows that a biological feature 'makes sense' in that it is an appropriate means to an end. This renders the feature in question intelligible in a way that non-adaptationist explanation cannot. Similarly, explaining a social phenomenon by tracing it to the deliberate actions of individuals renders the phenomenon intelligible in a particular way, in terms of the actors' reasons. For Weber, this type of subjective understanding (*verstehen*) marks the divide between the natural and social sciences. But we have seen that adaptationist explanation yields a type of understanding that is in many ways more akin to what we get from intentional-psychological than from physical explanation.

The agential idiom can help elucidate this parallel. Expressed in agential terms, Williams' point is that in biology, it is individuals rather than groups that are usually agent-like, that is, have goals towards which their phenotypic traits conduce. The typical biological group is akin to a collection of agents with different goals, not a collective agent with a single goal, on his view. When put this way, Williams' point is seen to be a close relative of Weber's. Both involve the claim that certain collective entities—biological and social groups, respectively—are not agent-like, and thus should not be appealed to in the respective scientific explanations. Social groups are not agent-like because they do not perform actions; biological groups are not agent-like because they lack the required unity-of-purpose.

The parallel holds in two further respects. Firstly, Williams realized that individual and group interests are not necessarily aligned; thus free-riders would tend to undermine the functionality of a biological group. This mirrors Elster's objection to Marxism. Just as the interests of an individual capitalist need not coincide with those of capitalists as a whole, so the evolutionary interests of an individual organism need not coincide with those of their group or species. Explaining a social phenomenon in terms of class interests is thus mistaken for Elster, as is explaining a biological phenomenon in terms of group evolutionary interests, for Williams. In both cases, failure to adhere to individualist strictures is a potential source of error.

Second, Williams' idea that group properties are often byproducts of individual adaptations echoes a point made by von Hayek (1942), who argued that the actions of many individuals could lead to unintended social phenomena of considerable complexity, for example, markets. Just as no individual consciously intends their action to lead to a well-functioning market, biological individuals do not usually 'intend' to produce group-beneficial outcomes, in the sense that doing so is not their evolutionary purpose. In both cases, the locus of agency and purpose is at the individual level; group features are to be explained as unintended side-effects.

2.4.1 Beyond individualism?

If this reconstruction of the dialectic is correct, it follows that there are two ways in which the demand for individualistic explanation might be resisted, in both biology and social science. First, one might hold that the characteristic mode of explanation—adaptationist or intentional-psychological—is not appropriate for the phenomenon in question. Second, one might dispute the claim that the relevant properties–possessing adaptations or having intentional states–are solely possessed by individuals.

Arguments of both sorts are found in both cases. In social science, Durkheim's opposition to individualism stemmed precisely from his rejecting the claim that social phenomena could only be explained in terms of individual motives; 'a social fact can only be explained by another social fact', he claimed (1895, p. 145). In biology, opponents of adaptationism argue that explanations in terms of developmental or phylogenetic constraint can sometimes provide more illumination than adaptationist explanations, or at least are a key part of the jigsaw (e.g. Amundson 2005).

However, it is the second way of opposing individualism that is of more interest here. In social science, this finds expression in the idea that collective entities such as committees and firms do sometimes qualify as agents that make choices and perform actions, *contra* Weber. The parallel in biology is the idea that certain biological groups are in fact adapted units with features that evolved because they are group-beneficial, *contra* Williams; this is what proponents of the superorganism concept hold and is what motivates agential thinking as applied to groups, discussed in section 2.3. In both cases, the claim is that groups as well as individuals can be agent-like in the relevant respects, at least in certain cases.

List and Pettit (2011) offer a detailed defence of group agency in social science. They argue that some social groups 'are agents in their own right with minds of their own' (p. 77). Groups can have mental states (or 'attitudes') such as beliefs, desires, and preferences which explain the group's actions—as when we say that the committee recommended an increase in fuel tax because it believed this was the fairest solution. List and Pettit describe their view as 'individualist'; but by this they simply mean that a group's properties, including its mental states, supervene on (i.e. are fully determined by) the properties of its constituent individuals. When it comes to explanation rather than ontology, their view is clearly anti-individualist, since they hold that some social phenomena can be explained in terms of group beliefs and desires, and that such explanations are irreducible to individual-level explanations.

The List and Pettit theory has one interesting point of contact with the foregoing biological discussion. Much of their focus is on how a group's beliefs and desires (or preferences) relate to those of its constituent individuals. The point here is that the individuals within any group may well believe and want different things, so it is unclear whether a unified 'group belief' and 'group preference' will even exist, and it is also unclear how to construct it. List and Pettit offer a precise account of the conditions under which individual beliefs and preferences can be aggregated to form group-level beliefs and preferences; this is necessary for a group to be a rational agent, they argue.

To see the biological connection, recall our unity-of-purpose constraint as applied to groups. For a biological group to be agent-like, its evolved traits must have a common goal: enhancing the group's fitness; otherwise we cannot talk sensibly of what a biological group 'wants' or is 'trying' to do. This in turn requires that the evolutionary interests of group members be aligned (or nearly enough), as otherwise they will pursue different goals and internal conflict will ensue. We saw that if individuals' interests are aligned, then their preferences over the relevant feasible set will be identical, and the group's preference may simply be equated with the individuals' preferences.

In effect, this means that the conceptual problem for group agency on which List and Pettit focus—how to aggregate individual attitudes into a group attitude—is automatically solved in the evolutionary biological context. More precisely, if a biological group qualifies as an agent in the sense relevant for adaptationist theorizing, that is, if the unity-of-purpose constraint is satisfied, then the aggregation problem will necessarily have a trivial solution. Therefore the biological mechanisms that serve to align the interests of group members—such as policing and clonality—thereby solve the evolutionary analogue of the aggregation problem that List and Pettit identify as the chief stumbling block for the notion of group agency.

However, in one respect the List and Pettit analysis of group agency is more liberal than our biological analysis. They allow a group to count as an agent despite substantial disagreement (divergence of attitude) among group members, as long as the aggregation problem can be solved; but a biological group, we have argued, is only agent-like to the extent that its members 'agree', that is, have common evolutionary interests. This difference is entirely appropriate. A committee whose members disagree may be able to speak with a common voice if a suitable compromise is found; but a biological group riven by internal conflicts of interests cannot negotiate its way to a solution, or find a compromise by which all parties agree to abide. So for human social groups, greater internal disagreement is compatible with group agency than in the evolutionary case.

2.5 The Biological Veil-of-ignorance

We know that for a biological group to behave as an adapted unit, and thus be agent-like, the interests of its constituent individuals must somehow be aligned; otherwise they will pursue their own goals, not a common group goal. One way that this alignment can be achieved is if individuals are deprived of information, or forced behind a 'veil-of-ignorance'. The ensuing inability to discriminate between possibilities can restrict individuals' ability to pursue goals that conflict with their group's welfare, ensuring they work for the common good.[13]

To illustrate, consider an example used by Queller and Strassmann (2013). In many ant species, new colonies are founded by small groups of unrelated queens. They cooperate with other until the first cohort of workers is produced, working together for a common goal; but they then fight to the death until only one queen remains. Clearly if a queen knew that they would be one of the losers in this fight, they would eschew the cooperative effort and instead take their chances on founding a new colony alone. However, queens do not know this; so joining the group makes sense if the average payoff from joining exceeds that from going it alone. The veil-of-ignorance thus aligns the interests of the queens, facilitating cooperation where it would not otherwise evolve.

A similar point applies to genes within organisms, as Ridley (2000) notes. In sexual species, the genes within a single organism do not have perfectly aligned interests, for they are not transmitted as a unit. This opens the possibility that a gene could profit at the expense of its host organism, for example, if it can preferentially assist copies of itself in the gametes, or in other organisms. For example, if a gene 'knows' that it was paternally inherited, then it has no interest in the future reproduction of its host organism's mother, so will try to extract as many resources from her as possible; there is empirical evidence that this actually happens. Or to take a different example, if a gene in a gamete 'knows' that another gamete does not contain a copy of

[13] See Frank (2003), Ridley (2000), Okasha (2012), and Queller and Strassman (2013) for discussions of the veil-of-ignorance notion in biology.

itself, it will profit from killing it, to the detriment of the organism.[14] Unsurprisingly, organisms have evolved ways of suppressing or scrambling this information, such as recombination, which deprives genes of the ability to enhance their own interests at the collective's expense. Again, a veil-of-ignorance serves to align otherwise divergent interests.

The veil-of-ignorance concept has it origins in economics and political philosophy, not biology. Both Harsanyi (1953, 1955) and Rawls (1971) asked how a rational agent would choose between alternative social or political arrangements if they lack information about which member of society they will be. An agent in this epistemic position cannot choose in a self-serving way, so occupies the role of 'impartial observer'. For Harsanyi and Rawls, the veil-of-ignorance was thus a thought experiment with which to address questions of social and economic justice. Rawls maintained, controversially, that from behind the veil an agent would choose the social alternative that maximizes the welfare of the least well-off member of society. By contrast, Harsanyi argued that the chosen alternative would be utilitarian, that is, would maximize society's total welfare.

How closely does the biological veil-of-ignorance map onto the original Harsanyi/Rawls idea? Clearly there are differences, but there is also a common structure. For the ethical issue these authors address arises because the individuals in society have different interests—they disagree on which social alternative is best. However, from behind the veil, any individual would choose in the same way, on both Rawls' and Harsanyi's theories, a point that Rawls himself emphasized.[15] So one effect of the veil is to reconcile the individuals in society with each other, allowing them to reach agreement. Similarly, the biological veil-of-ignorance reconciles the interests of the individuals, or genes, in a single collective, allowing them to 'agree' on what to do.

In fact, I think the veil-of-ignorance concept actually works better in biology that in its original setting, for the following reason. When the Harsanyi/Rawls argument is spelled out precisely, it is seen to rest on assumptions that are hard to justify. However, in the biological case the corresponding assumptions are innocuous, and the veil-of-ignorance argument leads directly to a utilitarian conclusion: individuals (or genes) behind a veil will seek to maximize the average or total fitness of their collective. These points are argued below.

[14] This occurs in some cases of male meiotic drive, in which a pair of linked genes on the same chromosome form a 'poison-antidote' pair, one producing a toxin which kills gametes that don't have the antidote. Recombination breaks the linkage, ensuring that a poison gene doesn't 'know' whether a potential victim contains a copy of itself or not (Haig and Grafen 1991, Ridley 2000).

[15] 'Since the differences among the parties are unknown to them, and everyone is equally rational and similarly situated, each is convinced by the same arguments. Therefore, we can view the agreement in the original position from the standpoint of one person selected at random. If anyone after due reflection prefers a conception of justice to another, then they all do, and a unanimous agreement can be reached' (Rawls 1971, p. 120).

2.5.1 The Harsanyi-Rawls argument

To fix ideas, consider the following standard way of modelling the veil-of-ignorance idea in social choice theory. There is a finite set of individuals in society, and a finite set S of social alternatives. For example, an alternative could be a possible way of dividing society's resources among the individuals. We let $w_i(s)$ be a real number that measures the welfare of individual i in social alternative s. We assume that individuals prefer alternatives in which their welfare is higher. To each social alternative $s \in S$, we can associate a payoff vector $< w_1(s), \ldots, w_n(s) >$, recording each individual's welfare in that alternative, where n is the number of individuals. How should society rank the alternatives? There are many conceivable ways. One is to rank by average or total welfare across individuals, as per utilitarianism. Another is to rank by the welfare of the least well-off individual, as per Rawls's 'maximin' idea.

It is here that the veil-of-ignorance enters. The basic idea, as we have seen, is that society's ranking of the social alternatives should be equated with that of an impartial observer, who does not know which individual they will be; following Harsanyi, we may take this to mean that they have an equal chance of being anyone (or believe that they do).[16] It helps to think of the observer not as a real person, but as a role that any individual can imaginatively occupy. Without serious damage to the core idea, we may equate the observer's uncertainty about their future identity with uncertainty about their future payoff. Thus from behind the veil, the observer is in effect required to choose between lotteries (probability distributions over alternatives), rather than actual alternatives. To each alternative $s \in S$, there corresponds a unique equi-probable lottery which yields each component $w_i(s)$ of the payoff vector $< w_1(s), \ldots, w_n(s) >$ with probability $1/n$.

Now, *if* we assume that the observer evaluates lotteries by their expected payoff, then a utilitarian-like conclusion follows immediately. For the basic idea says that society should prefer social alternative s to t if and only if the observer prefers the equi-probable lottery that corresponds to s over the one that corresponds to t. Presuming the observer evaluates by expected payoff, this in turn holds if and only if:

$$\frac{1}{n}\sum_i w_i(s) > \frac{1}{n}\sum_i w_i(t)$$

that is, society's average (or total) welfare is higher in alternative s than t. This implies, of course, that the distribution of welfare among individuals is of no consequence; only the total amount matters.

This simple formulation of the veil-of-ignorance argument differs from both Rawls' and Harsanyi's, though arguably captures the core idea. Rawls rejected the equation of ignorance of one's future position in society with having an equal chance of being anyone, so objected to the probabilistic formulation above; he also argued, notoriously,

[16] Here Harsanyi invokes the 'principle of indifference' of classical probability theory.

that a rational decision-maker from behind the veil will use an unorthodox decision rule, which attends only to their payoff under the worst possible outcome. Both points have been extensively debated in the literature; I share the standard view that Rawls' arguments here are unsatisfactory.[17]

Compared to Harsanyi, the key difference is that our formulation takes as primitive the notion of 'welfare'; and it implicitly assumes that welfare is measurable on an interval (or cardinal) scale and is interpersonally comparable, for otherwise it is not meaningful to compute expected welfare, nor to add individual welfare values. These are rather weighty assumptions. Harsanyi's own formulation was in terms of *utility* not welfare; but he did not take the utility notion as primitive. Rather, preference was his primitive notion, as per revealed preference theory (see section 1.7). He then introduces utility by assuming that individuals, and the observer, obey the axioms of expected utility theory, which implies that their preferences can be represented by a real-valued, cardinal utility function. This then supplies a justification for the key assumption that, from behind the veil, a lottery is evaluated in accordance with its expected payoff. Given this set-up, Harsanyi then shows that society's utility, in any alternative, can be expressed as the average of the individuals' utilities (if all the utility functions are suitably chosen).

Harsanyi's attempt to deduce utilitarian ethics from expected utility theory has met with much criticism. Indeed there is now a consensus that it cannot be done.[18] The basic problem is that utility, as it features in expected utility theory, is not interpersonally comparable—each individual may re-scale their utility function as they please. Thus it is essential to use a non-preference-based concept of utility or well-being, if the veil-of-ignorance argument is to yield a meaningful utilitarian conclusion, as opposed to a mere utilitarian-like formula, as Weymark (1991) and Mongin (2001) stress. This is why we took 'welfare' as primitive above, and simply posited that individuals prefer alternatives in which their welfare is higher. But even if this starting point is granted, one then needs to explain why, from behind the veil, a lottery should be evaluated by expected payoff. This does not follow from decision-theoretic principles, as it would do if we were talking about utility.

In short, Rawls' version of the argument rests on an eccentric decision-theory, while Harsanyi's faces a dilemma. If 'welfare' means utility in the sense of standard decision theory, then it is clear why evaluation from behind the veil should be by expected payoff, but it makes no sense to add individuals' utilities. If 'welfare' means something else, then even if it is cardinally measurable and interpersonally comparable, it is unclear why lotteries should be evaluated by expected payoff.

In a biological setting, none of these problems arise, for the relevant payoffs are in units of biological fitness, rather than utility or welfare. In a given social alternative, each individual receives a particular fitness value, that is, number of offspring. We can

[17] See Sen (2009) for a telling critique.
[18] See in particular Sen (1976, 1977a), Weymark (1991), and Mongin (2001).

sensibly talk about which alternative an individual 'prefers', but this is derivative from their fitness function, not vice-versa, as noted in section 1.7. It is unproblematic to add fitness values across individuals in a group. And if an individual faces a lottery over social alternatives, that is, if its number of offspring is a random variable, it is entirely reasonable to assume that the lottery's expected value is the quantity that the individual will care about (or appear to care about); this is the standard way of measuring biological fitness when individuals face demographic uncertainty.[19] Thus our veil-of-ignorance model can be given a biological interpretation, free from the conceptual problems that plague the Harsanyi/Rawls original.

2.5.2 *The ant colony example*

To illustrate, consider again the ant colony example. We may take each social alternative in S to specify the colony's total reproductive output, as measured by the number of offspring colonies, and how this output is divided between the n foundresses. Since all but one foundress gets killed, we need only consider alternatives in which the total output goes to a single queen, that is, where the payoff vector $< w_1(s), \ldots, w_n(s) >$ contains only one non-zero component. Thus for example, $< 20, 0, \ldots, 0 >$ is the alternative in which foundress 1 is the sole survivor and the colony produces a total of 20 offspring colonies. We assume that total colony output has a maximum, so finiteness of S is assured.

Clearly, without a veil-of-ignorance, the foundresses' preferences over the set S differ. Each prefers an alternative in which they are the sole survivor to one in which they are not, irrespective of total colony output. Thus consider two alternatives, $s = < 20, 0, \ldots, 0 >$ and $t = < 0, 10, \ldots, 0 >$. Clearly, foundress 1 prefers s to t while foundress 2 has the opposite preference. So interests are not aligned. The queens 'disagree' about which social alternative is best so will not act with a common purpose; rather, each will try to bring about the alternative in which it does best.

However, from behind a veil-of-ignorance, matters are different. If colonies are founded by n queens, then any queen has probability $1/n$ of being the sole survivor. So if queens lack information about who will win the fight, then in effect they need to evaluate equi-probable lotteries, rather than alternatives. If the queens perform actions that eventuate in social alternative s, then each has probability $1/n$ of receiving the total payoff in s, that is, $\sum_i w_i(s)$. Therefore each queen, if her behaviour has been optimized by natural selection, will perform actions that conduce towards the same goal, namely maximization of the total colony output. The fact that the total output is distributed unequally, all going to one queen, makes no difference to a queen's decision, given her ignorance. That is, from the behind the veil-of-ignorance, each queen behaves as if she were a utilitarian.

[19] Demographic uncertainty means that the risk is uncorrelated across organisms, that is, each faces an independent lottery; the contrast is with environmental uncertainty; see section 8.3.2.

The reason that the veil-of-ignorance argument works in this biological context is simple. The prerequisites necessary for the argument to work—payoffs that are interpersonally comparable, and a justification for assuming that lotteries are evaluated by expected payoff—are satisfied immediately if payoffs are in units of biological fitness. In particular, the second assumption is justified by the basic logic of natural selection itself. In general, if an individual with a given trait does not receive a definite fitness payoff, but rather a probability distribution (lottery) over possible payoffs, then it is the expected payoff that determines the trait's evolutionary success—as long as the probability distribution is independent across individuals. (The need for this independence assumption is explained in section 8.3.2.) So the justification for assuming that, from behind the veil, a well-adapted individual will respond to uncertainty by evaluating a lottery in terms of expected fitness is immediate, in a biological context.

I find it quite striking that the veil-of-ignorance argument, which is fraught with conceptual problems in its original setting, applies so neatly in biology. It appears that an analogue of the Harsanyi/Rawls thought experiment has actually been put into practice by evolution, and has resulted in cooperative outcomes where they would not otherwise have arisen. It is an open question whether all the putative biological examples of the veil-of-ignorance have the structure of the above example; I do not want to take a stand on this here. For the moment, the point is just to illustrate how, in a rather unexpected context, a model from rational choice theory can illuminate biological phenomena.

2.6 Conclusion

In some applications of agential thinking, genes and groups, rather than individual organisms, are the entities that are treated as agent-like. Applied to genes, this allows us to make sense of intra-genomic conflict, by yielding adaptationist explanations of its phenotypic consequences; this is the true merit of the 'genes as agents' concept, and the only circumstance in which the gene *must* be taken as the agent, in order to apply agential thinking. The circumstances in which groups may be treated as agents are relatively infrequent, for the necessary unity-of-purpose is often lacking, owing to internal conflict. There is an instructive parallel here with group agency in social science; for in both the biological and social-scientific cases, the key issue is how the divergent interests of group members may be aligned. The overall moral is that applying agential thinking to genes and groups can be heuristically valuable, but only under fairly specific conditions.

PART II

The 'Goal' of Fitness Maximization

3

Wright's Adaptive Landscape, Fisher's Fundamental Theorem

3.1 Introduction

The issue of fitness maximization, or optimization, has long been a source of controversy in biology. Optimization, in its standard mathematical sense, refers to the choice of a best element from a given set by some criterion. Intuitively both natural selection and adaptation have *something* to do with optimization, and indeed some evolutionists regard this as trivially true.[1] After all, natural selection involves fitter variants being chosen at the expense of the less fit; while adaptation means that an organism's evolved traits enable it to survive and reproduce in its environment, which implies that alternative traits would achieve this less well. So in a sense, optimization seems implicated whenever selection occurs, and whenever an organism evolves adaptations to its environment.

Why then is fitness optimization a matter of controversy? I think we can distinguish three reasons: one that pertains to selection, the second to adaptation, and the third to the relation between them. First, the sense in which selection is obviously an optimizing process is not what critics of optimization usually mean. Although at any point in time, natural selection chooses between variants according to fitness, it does not follow that over time this process will lead to maximization of any quantity, in part because the selective environment may itself change. Second, although adaptation means that an organism's traits are optimal relative to some set of alternatives, this set may not be especially large, owing to constraints, and will invariably be smaller than the set of all conceivable alternatives; thus the use of 'optimality models' to predict evolved trait-values may mislead. Third, it is not clear that natural selection will always lead to adaptation anyway, nor even that it will 'tend' to do so, owing to complexities such as inheritance, frequency-dependence, and stochasticity. The simple view of Darwinian evolution as leading to fixation of the best variants in a population, as judged by some fixed standard, is naive.

This chapter and the two that follow examine a number of prominent ideas about evolution and fitness-maximization. The focus in this chapter is on the adaptive

[1] Thus Mayr (1998) writes: 'to be sure, evolution is an optimization process' (p. 104).

landscape concept originally due to Sewall Wright, and the 'fundamental theorem of natural selection' (FTNS) of R. A. Fisher. Chapter 4 focuses on Alan Grafen's 'maximizing agent analogy', and looks at adaptive dynamics, a framework for analysing frequency-dependent selection. Chapter 5 focuses on social evolution, in particular on W. D. Hamilton's idea of inclusive fitness maximization. Some of these ideas are decades old, while others are of more recent pedigree; all are controversial.

I focus on these ideas partly because of their intrinsic interest, partly because they raise delicate conceptual issues, and partly because they relate, with varying degrees of directness, to agential thinking in biology. The notion that natural selection pushes populations up adaptive peaks suggests a quasi-teleological view of evolution, as a goal-directed process; this ties in with what we called agential thinking (type 2), in which the process of natural selection is analogized to an agent's deliberate choice. Grafen's and Hamilton's ideas, by contrast, tie in with agential thinking (type 1), since they concern the idea that evolved organisms will behave as if they were trying to maximize some quantity, that is, like rational agents. Thus by studying the debates over fitness-maximization in biology, we can shed light on both types of agential thinking, and the relation between them.

The structure of this chapter is as follows. Section 3.2 focuses on the adaptive landscape concept, explores its limitations, and asks why it continues to polarize modern evolutionists. Section 3.3 analyses Fisher's FTNS, expounds its derivation in a simple case, and considers its interpretation. Section 3.4 concludes.

3.2 The Adaptive Landscape

[Lack of] formal proof that populations do climb adaptive peaks has never... dissuaded evolutionary biologists from taking the ascent for granted. P. O'Donald (2000)

The adaptive landscape was invented by Wright (1932) as a graphical tool to help understand the dynamics of an evolving population. Wright's focus was on Mendelian populations evolving under the influence of natural selection and other evolutionary forces, such as random drift. Wright had an elaborate theory, known as the 'shifting balance' theory, about how selection and drift would jointly influence a population's evolutionary trajectory, and devised simple diagrammatic aids to help visualize the process.[2]

The adaptive landscape, as I will use the term, is the graph of a function which relates the average fitness of a biological population to the frequency of the different genes within it, in a given environment. One might wonder why such a function exists—surely a population's average fitness really depends on its entire distribution of genotypes? However, given certain assumptions, including random mating, this

[2] For discussions of the shifting balance theory, see Coyne *et al.* (1997), Goodnight (2012), and Wade (2012).

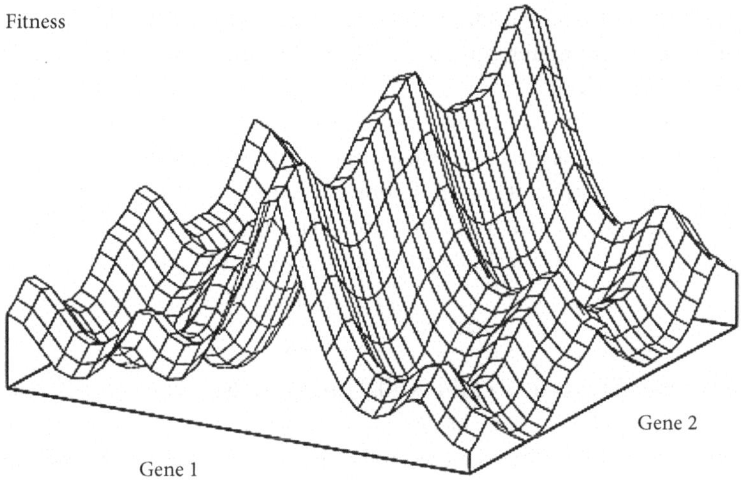

Figure 3.1. Adaptive landscape

distribution will itself depend only on the gene frequencies.[3] The adaptive landscape is highly multi-dimensional, with a separate axis for each gene in the population, but three-dimensional slices are easily visualized. Thus Figure 3.1 depicts a population's average fitness as a function of the frequency of two genes (at different loci). Peaks on the landscape correspond to high fitness, troughs to low fitness. A population is represented by a single point in the landscape and its evolution as movement through the landscape.

Wright also described a quite different landscape, plotting the fitness of an individual organism, rather than a population, as a function of the frequency of each gene within its own genotype ($= 0, \frac{1}{2}$, or 1 for a diploid species). On such a landscape, a population would be represented by a cloud of points rather than a single point (Provine 2001). The distinction between the two sorts of landscape is important, not least because a gene's frequency within an individual is a discrete quantity, so cannot vary continuously as the population frequency can, and as Wright's diagrams depict. Following modern usage, I use 'adaptive landscape' to refer to the plot of population mean fitness against population gene frequency, and 'individual fitness surface' to refer to the plot of individual fitness against individual genotype or phenotype.

Wright's guiding idea was that natural selection would tend to push a population to the top of the nearest peak in the adaptive landscape. Thus evolution by natural selection is a hill-climbing process, always changing gene frequencies so as to increase average fitness. This was nicely captured by Lewontin (2000): 'mean fitness is like the altitude in a mountain range, and natural selection is like an inner compulsion of a

[3] The other key assumption is linkage equilibrium, that is, absence of statistical association between genes at different loci.

climber to climb yet higher and higher. Hence evolution by natural selection is seen as a maximizing or optimizing process' (p. 321). Note that Lewontin is describing the idea here, not endorsing it.

Wright's view of evolution as hill-climbing stemmed from his analysis of a simple evolutionary model, namely selection at a single locus in a random mating population with non-overlapping generations and constant genotype fitnesses. If there are two alleles *A* and *B* at a locus in such a population, Wright (1932) showed that the change in frequency of allele *A* over a single generation may be written as:

$$\Delta p = \frac{p(1-p)}{2\overline{w}} \frac{d\overline{w}}{dp} \tag{3.1}$$

where p is the initial frequency of the *A* allele and $\overline{w}$ is the average population fitness. Note that the term $p(1-p)/2$ is the genetic variance at the locus in question.

In this diallelic example, the adaptive landscape becomes a simple two-dimensional plot of $\overline{w}$ against p; the slope at any point is $\frac{d\overline{w}}{dp}$—which is the second term on the right-hand side of equation (3.1). Since the other right-hand side term is necessarily non-negative, the sign of this slope determines whether the *A* allele will spread. Therefore starting from any intermediate value of p, natural selection will change p so as to increase $\overline{w}$, that is, the population will ascend the landscape. As long as the genetic variation is not exhausted, gene frequencies will continue to change until $\overline{w}$ is at a local maximum.

The idea that selection will drive a population up a mean fitness gradient is intuitive, and might even be thought tautological. After all, natural selection involves the fittest types proliferating at the expense of the less fit; so it stands to reason, surely, that the average fitness of a population after an episode of selection will be greater than before? However, this intuitive argument ignores reproduction and inheritance. An organism does not transmit its fitness value to its offspring directly; rather, in a sexual population, it contributes a random sample of its genes, which together with genes from other organisms determine the genotypes of its offspring, which in turn determine the offspring's fitness (given the environment). So even when genotype fitnesses are fixed, it is not obvious that natural selection must cause $\overline{w}$ to rise.

Rice (2004) describes Wright's equation (3.1) as 'one rigorous formulation of the intuitive idea of adaptation', since it shows that 'organisms evolve traits that maximize the ability of a population of those organisms to increase in size in its particular environment' (p. 11). (Note that the population's instantaneous growth rate is $\ln(\overline{w})$, so maximizing $\overline{w}$ is equivalent to maximizing the growth rate.) On this view, Wright's equation formalizes, in a Mendelian setting, Darwin's verbal argument that natural selection will lead organisms to evolve adaptations. I think that this accurately reflects how Wright and others understood the adaptive landscape; however, Rice's interpretation has been contested, as we shall see.

One further aspect of Wright's (early) view deserves mention. He argued that $\overline{w}$ plays a role akin to that of a 'potential function' in classical mechanics, that is, a function whose gradient determines the direction and magnitude of a dynamical system's change.[4] If true, this means that an evolving population not merely always moves uphill in the adaptive landscape, but *that it takes the steepest path*. So the population's trajectory would be deducible from the landscape's topography. Evolution by natural selection would then appear like a goal-directed process, mimicking what a conscious agent, trying to improve a population by the most efficient means possible, would do. In the diallelic case described by equation (3.1) the issue of whether the ascent is by the steepest path does not arise, for the landscape has a single dimension, that is, there is one way to the top. But with more than two alleles the issue does arise.

Wright's version of the adaptive landscape is not the only one. An alternative is to plot the landscape in phenotypic rather than genotypic space, following Simpson (1953). Phenotypic landscapes are widely used in quantitative genetics (which studies continuous traits, such as height, that are affected by many genes of small effect). In the canonical modern treatment, each horizontal axis denotes the population average of a given trait $\overline{z}$, while the vertical axis denotes mean population fitness $\overline{w}$, just as in Wright's version (Lande 1979). So if we consider n different traits, a population's phenotypic composition is described by the vector of trait means $\overline{\mathbf{z}} = (\overline{z_1}, \cdots \overline{z_n})$, represented by a point on an n-dimensional landscape; and phenotypic evolution is represented by movement on this landscape. The hill-climbing property is then the idea that natural selection will lead $\overline{\mathbf{z}}$ to evolve so as to maximize $\overline{w}$.

As with the genotypic version, the adaptive landscape in phenotype space derives from an underlying individual fitness surface, which in this case plots an individual's fitness w_i as a function of its own n-dimensional phenotype $(z_1, \cdots z_n)$. Again, one might wonder why a population's mean fitness can be written as a function of the trait means alone—surely it may depend on the entire distribution of phenotypes in the population? However, standard quantitative genetics relies on a particular assumption about the shape of the phenotype distribution, namely that it is multi-variate normal (Lande 1979). (The multi-variate normal distribution generalizes the familiar univariate normal distribution to more than one variable.[5]) This ensures that $\overline{w}$ depends only on the vector of trait means $\overline{\mathbf{z}}$.

The adaptive landscape has had a tangible influence on evolutionary biology—evidenced for example by the title of Dawkins' book *Climbing Mount Improbable*. But it has also been criticized, and there persists a sharp disagreement among biologists over the value of the landscape concept. Thus Svensson and Calsbeek (2012) describe it as 'a highly successful metaphor and scientific concept' (p. 301), while O'Donald

[4] Any conservative force field, such as a gravitational field, has an associated potential function, which is maximized at equilibrium. In later work, Wright (1949) retracted his claim that the population would evolve by the steepest path in the landscape, as Edwards (1994) notes.

[5] See Arnold *et al.* (2001) or Rice (2004) for explanation and discussion of this assumption.

(2000) describes the idea that populations climb adaptive peaks as a 'seductive and erroneous metaphor of the evolutionary process' (p. 621). Here I focus on three questions. How often does the hill-climbing property actually obtain? When it does obtain, what does this show? And when it fails to obtain, can this be interpreted as due to 'extraneous factors' other than selection itself?

3.2.1 Hill-climbing: myth or reality?

The first question has been extensively studied in formal evolutionary models, both genotypic and phenotypic. Let us start with the former. Wright's equation (3.1) shows that the hill-climbing property obtains for two alleles at a single locus, given fixed genotype fitnesses and random mating. If instead there are multiple alleles at the locus then hill-climbing still occurs, but the population does not climb the landscape via the steepest path; rather it traces a curved path to the summit (Edwards 1994, Rice 2004). If the random mating assumption is dropped, then hill-climbing need not occur; this was shown by an example of Fisher (1941) involving inbreeding, in which selection changes gene frequencies but $\overline{w}$ does not rise. If multiple loci are considered, then even with random mating and fixed genotype fitnesses, natural selection can lead $\overline{w}$ to fall; this was shown by an example of Moran (1964) involving a combination of epistasis (non-additivity of fitness) and recombination between loci. And of course, if genotype fitnesses are not fixed across generations, for example, because they are frequency-dependent, then $\overline{w}$ will not necessarily rise. This is fairly obvious: selecting the fittest variants need not increase mean fitness if the fitness of a variant changes as the population evolves.

These well-known points suggest that the hill-climbing property obtains only in special cases, and the steepest ascent property yet more rarely; so they are not general properties of biological evolution. Wright himself was aware that $\overline{w}$ would not always increase, in particular where genotype fitnesses were not fixed; but he hoped that an alternative quantity could be found that would be maximized by natural selection. However, the hunt for such a quantity has not been a success (Arnold *et al.* 2001). Indeed it is perfectly possible for selection to lead gene frequencies to cycle indefinitely, never attaining equilibrium; in this case it is obvious that there is no quantity that selection maximizes (Rice 2004). In its genotypic version, therefore, the notion that natural selection pushes populations up adaptive landscapes, for all its intuitive appeal, lacks theoretical support.

In its phenotypic version the situation is somewhat different. Standard quantitative genetics does recover the hill-climbing property for phenotypic evolution, given the specific assumptions on which it rests. The key assumptions are that an individual's fitness depends only on its own trait values, the phenotypic distribution is multivariate normal, and selection is weak (Lande 1979). (In effect, these assumptions ensure the absence of the genetic factors that can prevent $\overline{w}$ from being maximized, such as non-random mating and epistasis.) However, even then, the steepest ascent property is only satisfied in a special case. In general, as the vector of trait means

$\bar{\mathbf{z}} = (\overline{z_1}, \cdots \overline{z_n})$ evolves by natural selection, the population moves up the adaptive landscape but *not* by the steepest route (Lande 1979, Arnold *et al.* 2001). This point is worth explaining briefly.

The direction of steepest ascent on a phenotypic landscape is given by the gradient vector:

$$\nabla \overline{w} = \left(\frac{\partial \overline{w}}{\partial z_1}, \cdots, \frac{\partial \overline{w}}{\partial z_n} \right)$$

where each partial derivative is the slope of the landscape in the direction of trait z_i (evaluated at the population mean trait). This vector is orthogonal to the level set on which the population is located—just as the steepest path up a mountain, from a point on an ordinary topographical map, is to move at right angles to the tangent to the contour line. To see that the population need not evolve in the direction of $\nabla \overline{w}$, note that if a given trait z_i exhibits very little variance, selection cannot change its mean value by much, even if the value of $\frac{\partial \overline{w}}{\partial z_i}$ is substantial.

In general, the change in the vector of trait means $\Delta \bar{\mathbf{z}}$ is given by the product of the gradient vector $\nabla \overline{w}$ with the 'genetic variance-covariance matrix', denoted **G**. This is is an n x n matrix in which element (i, j) is the covariance between the additive genetic value of traits z_i and z_j, respectively.[6] This yields:

$$\Delta \bar{\mathbf{z}} = \frac{1}{\overline{w}} (\nabla \overline{w} \cdot \mathbf{G}) \tag{3.2}$$

which is a version of a classic equation in quantitative genetics (Lande 1979). Note the close analogy between equation (3.2) and Wright's original equation (3.1).

It is clear from equation (3.2) that the steepest ascent property does not usually obtain. For multiplying the gradient vector $\nabla \overline{w}$ by the **G** matrix yields a new vector which will point in a different direction—unless **G** is the identity matrix (or some scalar multiple of it). Therefore, only if all n traits have identical variance, and there is no covariance between any of them, will the population ascend the landscape by the steepest path. Empirically this is unlikely to be true, since many phenotypic traits, such as height and weight, exhibit significant correlations. The effect of such correlations is to deflect the population from the path of steepest ascent: it traces a curved path to the top of the landscape rather than going straight up.

The quantitative-genetic analysis of phenotypic evolution, enshrined in equation (3.2), partially vindicates the adaptive landscape view of evolution, by showing that the hill-climbing property holds in an important class of evolutionary models. This suggests that critics who dismiss the adaptive landscape as 'mere metaphor' have overstated their case (e.g. Kaplan 2008). However, restrictive assumptions have been made. Frequency and density-dependence have been excluded by assuming that an individual's fitness depends only on its own trait values. Moreover, the (multi-variate)

[6] The additive genetic value of a trait measures the value of a trait as predicted by a best-fit linear regression on the genes that underlie it. See Falconer and McKay (1996), or section 3.3.

normality of the phenotype distribution is critical to the derivation of the hill-climbing property. It is a reasonable idealization, for many phenotypic traits show an approximately normal distribution, but is mainly made to keep the mathematics tractable. Therefore, we should not conclude that increasing $\overline{w}$ is a general property of Darwinian evolution.

3.2.2 Hill-climbing and adaptation

When the hill-climbing property does obtain, what does this show? We saw that Wright's equation (3.1) has been viewed as a statement about adaptation. However, against this, it is sometimes argued that the sort of optimization that the equation describes, in which a property of a population, $\overline{w}$, is shown to be maximized at equilibrium, is fundamentally different from the Darwinian idea that individuals will evolve traits that maximize their fitness, which is the standard notion of adaptation. Both Grafen (2006) and Birch (2016) insist on this distinction. Similarly, Fisher (1941) wrote: 'objection should be taken to Wright's equation (3.1) principally because it represents natural selection, which in reality acts upon individuals, as though it were governed by the average condition of the species or inter-breeding group' (p. 58). Here Fisher implies that to construe Darwinian evolution as movement up an adaptive peak is to conflate individual with group advantage.

How should this issue be adjudicated? It is true that maximization of individual and population fitness are different notions, and that studying the latter is a rather indirect way of understanding the former. However, the mere fact that $\overline{w}$ is a population property does not preclude its evolutionary increase from telling us something about individual adaptation. For $\overline{w}$ is directly related to individual properties—genotype, phenotype, and fitness—which are the direct target of natural selection. For $\overline{w}$ to increase, then at least some individuals must evolve traits that make them fitter than their ancestors. Thus I see no objection on this score to regarding the hill-climbing property, when it obtains, as an expression of Darwin's idea that natural selection will optimize the fit of individuals to their environment—at least for the case of fixed genotype fitnesses. Indeed Fisher's objection is a curious one, since his FTNS also deals with a population-level property, and yet Fisher *did* interpret it as a statement about individual-level adaptation, as we shall see in section 3.3.

A quite different argument against equating an increase of $\overline{w}$ with individuals becoming better adapted is that the peak on the individual fitness surface need not correspond to the peak on the adaptive landscape (Gomulkiewicz 1998, Urban *et al.* 2013). This point is easily seen if we focus on a phenotypic landscape with a single continuous trait, normally distributed.[7] The top panel in Figure 3.2 depicts an individual's fitness w_i as a function of its own trait value z_i. The optimal trait value z^*

[7] The same point is made in relation to the genotypic landscape by Kirkpatrick and Rousset (2005); their example also invokes an asymmetric fitness function and a genetic constraint (heterozygote superiority) which prevents the optimal type from breeding true.

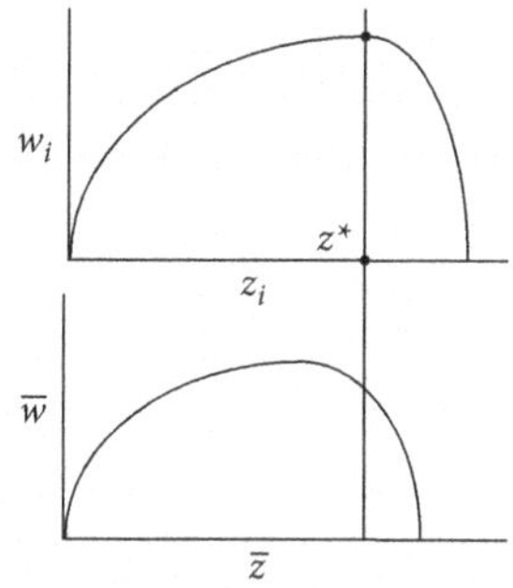

Figure 3.2. Individual versus population optimum

is intermediate, and the function exhibits a strong asymmetry: fitness falls away faster for $z_i > z^*$ than for $z_i < z^*$. The bottom panel depicts the corresponding adaptive landscape, which plots mean fitness $\overline{w}$ against mean trait $\overline{z}$. Note that the optimal mean trait value is less than z^*. This is because by assumption, there is a symmetric distribution of individual trait values around the population mean; so given the asymmetry of the individual fitness function, a population's fitness is maximized for a value of $\overline{z}$ less than the individual optimum z^*.

This *seems* to show that a population's ascending an adaptive landscape cannot always be interpreted as its constituent individuals becoming better adapted. However, note that the example depends essentially on the population not being monomorphic. If it is possible for all individuals to have exactly the same trait value, then the peaks on the adaptive landscape and individual fitness surface must align, irrespective of any asymmetry in the latter. But given the normal distribution assumption, individuals with the optimal trait value z^* cannot breed true—they beget offspring with a spread of trait values. Now inability of the optimal phenotype to breed true is a genetic constraint; and in the presence of such constraints, the idea that selection will lead individuals to evolve optimal phenotypes is untenable anyway, as is widely agreed. Therefore, the non-alignment of peaks does not refute the idea that hill-climbing captures selection's tendency to optimize individual phenotypes. Rather, it highlights the fact that constrained and unconstrained optimization may yield different outcomes.

If this is right, then hill-climbing on an adaptive landscape does capture natural selection's tendency to lead individuals to become better adapted, subject to the constraints that the landscape embodies. Thus evolutionary models in which the hill-climbing property obtains can be regarded as formal expressions of Darwin's argument that natural selection will tend to produce well-adapted individuals.

But the hill-climbing property is not universal, as we know. Does this mean that where it fails, selection does not produce adaptation? Not necessarily. It depends on why selection fails to increase $\overline{w}$. If it is because of frequency or density-dependence of fitness, the traits that evolve may nonetheless count as adaptations. In such cases, the trait that evolves will often (though not always) be an 'evolutionary stable strategy' (ESS); an individual with such a trait maximizes its fitness conditional on it being fixed in the population (section 4.3). Since an individual's environment may be taken to include the composition of its population, an ESS trait therefore does qualify as an adaptation, that is, it is fitness-maximizing given the environment; however, its evolution need not involve increase of $\overline{w}$. In such cases, failure of selection to increase $\overline{w}$ does not mean that selection is failing to produce adaptation; but in cases where genotype/phenotype fitnesses are fixed, it does mean this. That such cases are possible shows that selection is not theoretically guaranteed to produce adaptation, even if in practice it has often done so.

3.2.3 *Extraneous factors?*

This brings us to our third question: when the hill-climbing property does not obtain, can this be interpreted as due to extraneous factors, other than natural selection itself? The significance of this is that if the answer is yes, this would support the notion that natural selection has an inherent tendency to increase $\overline{w}$, as the counterexamples would be attributable to the intrusion of perturbing forces.

In some cases, the answer is plausibly yes. Where selection fails to increase $\overline{w}$ because of frequency or density-dependence, this can be attributed to the environment changing. It is obvious that if the external environment changes, for example, the climate, this may offset any selection-induced increase in a $\overline{w}$. And as noted above, an organism's selective environment includes not just external factors but also the genetic and phenotypic composition of the population to which it belongs. So changes to the latter, whether due to natural selection or other causes, imply that the relevant selective environment has changed—which may counteract selection's tendency to increase $\overline{w}$. The logic of this argument is explored further in section 3.3, in relation to Fisher.

However, in other cases where hill-climbing fails, invoking extraneous factors is less plausible. Non-random mating can prevent $\overline{w}$ from increasing, but there is no reason to treat random mating as the default condition from which to assess the 'unadulterated' effect of natural selection on $\overline{w}$; rather, it is one possible assumption about how diploid zygotes are formed from haploid gametes. It so happens that in a one-locus model, random mating yields the hill-climbing property; but this is a mathematical fact, not a demonstration that non-random mating involves a deviation from the intrinsic tendency of $\overline{w}$ to increase. The same is true of linkage disequilibrium (i.e. statistical associations between genes at different loci), which can lead $\overline{w}$ to decline in multi-locus models. Similarly, on the phenotypic landscape, failure of the standard quantitative-genetic assumptions can invalidate the hill-climbing property; again, there seems no good reason to attribute this to the presence of perturbing forces.

On the other hand, note that these failures of hill-climbing arise in the context of sexually reproducing populations. If reproduction is asexual, and if each individual passes on its traits faithfully to its offspring, and if an individual's fitness depends only on its own traits, then the hill-climbing property will necessarily obtain. For then, an individual's actual fitness value w_i will be transmitted faithfully to its offspring; from this it follows that $\overline{w}$ must increase as a result of natural selection. (This follows from the Price equation discussed in section 2.3.2.) Obviously no real population will satisfy the faithful transmission assumption. But it might be argued that this represents the purest imaginable form of Darwinian evolution, free from any extraneous factors including sexual mixing, and that hill-climbing is thus its default tendency.

There are two problems with this argument. Firstly, sexual reproduction is surely a structural feature of biological populations, not a perturbing factor. Second, the argument effectively reduces the idea that selection has an optimizing tendency to a triviality, and bypasses the reasons why the adaptive landscape is controversial. Natural selection's tendency to increase $\overline{w}$, construed this way, is comparable to the obvious fact that if you select the most expensive items in a shop, the average price of the items in your basket will exceed that of all the items in the shop. (This example is due to Michod (1999), who calls it 'everyday selection' (p. 60).) It may be that the appeal of the adaptive landscape stems partly from our tendency to assimilate evolution by natural selection, in biological populations subject to the complexities of biparental inheritance, to everyday selection. However, if so this is surely a mistake, not an indication that when evolution occurs in such populations, the inherent tendency towards hill-climbing is being frustrated by extraneous factors.

3.2.4 Why does the landscape metaphor persist?

Finally, I want to consider briefly a sociological question. Given that hill-climbing is not an inevitable property of Darwinian evolution, why is the adaptive landscape metaphor still so popular in biology? I can see four possible explanations. The first three are internal to the science, while the fourth is external.

The first is that models in which hill-climbing does occur are sufficiently important to justify continued use of the metaphor. The second is that the landscape metaphor is useful independently of whether the hill-climbing property obtains. The third is the empirical fact that organisms are often well-adapted to their environment. The fourth is that teleological modes of thinking have a peculiar grip on the human mind, including those of scientists.

I suspect that all four explanations play a role. Certainly the recovery of hill-climbing in the standard quantitative-genetic model of phenotypic evolution is part of the story (explanation 1); it is no accident that exponents of this model are among the most enthusiastic proponents of the landscape concept (e.g. Arnold *et al.* 2001). It is also true that adaptive landscapes have heuristic uses that don't depend on whether $\overline{w}$ is maximized (explanation 2). For example, Coyne *et al.* (1997) interpret the peaks in their adaptive landscape model as stable equilibria, remaining deliberately agnostic about whether they represent fitness maxima or not.

Explanation 3 is also a factor, I think. Hill-climbing may not be a universal property of evolutionary models; but given that well-adapted organisms have in fact evolved, in countless species, it is tempting to assume that a lot of hill-climbing must have occurred in the past. There is thus an empirical justification for assuming that selection pushes populations up adaptive peaks, namely that many populations seem to have arrived at them. I suspect that this reasoning informs many informal invocations of the adaptive landscape concept in biology. The reasoning is certainly plausible, at least for traits whose contribution to an organism's fitness is frequency-independent, though familiar caveats about the perils of uncritical adaptationism apply.

In addition to these 'rational' explanations, I suspect that another factor may also be at work, namely the pervasive tendency to view the evolutionary process in teleological and agential terms, discussed in chapter 1. (Recall Darwin's description of natural selection as 'daily and hourly scrutinizing, throughout the world, every variation . . . silently and insensibly working . . . at the improvement of each organic being' (1859, p. 133).) Of course modern biologists know well that evolution is a blind, mechanical process. However, metaphors can be powerful, and teleological thinking is deeply ingrained in the human psyche. The image of a population ascending an adaptive landscape fits naturally with the idea that Darwinian evolution has a *telos*, and with the use of agential and intentional idioms to describe it. Though I cannot prove it, I suspect that there is a two-way influence here: the hill-climbing view of evolution both reinforces and is reinforced by the metaphorical personification of natural selection as an agent with a goal ('mother nature'), which she pursues by choosing some variants over others.

Our discussion of the adaptive landscape shows why this metaphorical personification—what we called agential thinking (type 2)—is misleading. For the hill-climbing property is not universal, and even when it does obtain, the steepest ascent property generally does not. Populations typically trace a roundabout route to the top of the landscape rather than going straight up (contrary to what is often asserted).[8] So if mother nature's goal is to maximize a population's mean fitness by changing its genetic or phenotypic composition over time, she has chosen a rather inefficient means of achieving it.

3.3 Fisher's Fundamental Theorem

Wright and Fisher knew that selection need not maximize mean population fitness . . . but seemed concerned with phrasing [their] results in terms of something being maximized wherever possible.

S. Rice (2004), p. 36

[8] For example, a fairly recent paper in *Nature* states that 'under natural selection [a population] will tend to evolve along the steepest path uphill towards higher fitness . . . eventually moving the mean phenotype of a population to a local fitness maximum' (Dethlefsen *et al.* 2007, p. 817).

A quite different take on evolution and optimization comes from Fisher's fundamental theorem of natural selection (FTNS), first presented in his famous 1930 book. Fisher believed that his theorem expressed a deep truth about Darwinian evolution; it held, he said 'the supreme position among the biological sciences' (1930, p. 47). However, Fisher's own assessment of his theorem has not been universally accepted, and its significance continues to divide commentators today, in part because it raises tricky philosophical issues to do with causality. In this section I explain the theorem and its rival interpretations before offering my own assessment of what it shows.

Fisher's original statement of the FTNS was: 'the rate of increase of fitness of any organism at any time is equal to its genetic variance in fitness at that time' (1930, p. 46). By the 'fitness of an organism' Fisher meant the Malthusian growth rate[9] of an evolving population, which is the continuous-time analogue of average fitness $\overline{w}$; indeed in a later formulation Fisher replaces 'fitness of an organism' with 'average fitness of the population' (1941, p. 377). Fisher's treatment used a continuous-time model with overlapping generations; however, here I use the simpler discrete-time version of the FTNS with separate generations, following Ewens (1989, 2010). In place of the instantaneous rate of change, the discrete-time version deals with the change in mean fitness from one generation to another, or $\Delta\overline{w}$.

By the 'genetic variance in fitness' Fisher meant a certain portion of the total variance in fitness, namely that attributable to additive gene action; this is explained below. Since the variance of any variable is by definition non-negative, Fisher's statement *seems* to imply that the average fitness of an evolving population can never decline, which is how the FTNS was originally understood. If true, this implies that an evolving population will always move uphill in the adaptive landscape; and indeed Wright (1932) appealed to the FTNS in support of his hill-climbing view of evolution.

However, this was not what Fisher meant, as commentators have gradually come to realize.[10] Fisher was not concerned with the total change in $\overline{w}$, but rather with the 'partial change' that he regarded as directly attributable to the action of natural selection. In particular, Fisher was concerned to exclude any changes in $\overline{w}$ that are due to what he called 'environmental deterioration'—a notion that he understood rather broadly. When such environmental changes are excluded, the remainder is the change due to natural selection. Schematically, Fisher's conception was as follows:

$$\Delta\overline{w} = \Delta\overline{w}_{\mathrm{NS}} + \Delta\overline{w}_{\mathrm{E}}$$

where $\Delta\overline{w}$ is the total (or actual) change in average fitness from one generation to another; the components $\Delta\overline{w}_{\mathrm{NS}}$ and $\Delta\overline{w}_{\mathrm{E}}$ are the partial changes due to natural selection and the environment, respectively. The FTNS concerns only the partial

[9] The Malthusian growth rate is the term r in the exponential growth model $N_{t+1} = N_t e^r$, and thus equals $\ln(N_{t+1}/N_t)$. Note that (N_{t+1}/N_t) equals the population's average fitness $\overline{w}$ at time t.

[10] See Price (1972b), Ewens (1989), and in particular Edwards (1994) who offers a blow-by-blow account of how this gradual realization unfolded.

change $\Delta\overline{w}_{NS}$—this is the quantity that the theorem shows to be non-negative. However, since the environmental component of the change $\Delta\overline{w}_{E}$ may be negative, and often will be according to Fisher, nothing follows about the sign or magnitude of $\Delta\overline{w}$ itself. This is why the FTNS does not in fact vindicate the hill-climbing view of evolution, as Fisher (1941) emphasized.

Before delving into the technical details, the abstract logic of Fisher's argument is worth considering. His aim was to say something general and precise about how natural selection would affect an evolving population. Now clearly, a population's mean fitness may change for reasons that have nothing to do with natural selection. For example, if deforestation destroys a population's habitat, or if predators become more abundant, then $\overline{w}$ may decline. Given such possibilities, it seems unlikely that anything general can be said about the total change in $\overline{w}$ in real biological populations—it simply depends on too many things. Fisher thus wants to bracket off such environmental changes in order to isolate the effect of natural selection itself on $\overline{w}$. That is, he wants to consider the effect of natural selection in a constant environment.

Expressed this way, Fisher's logic may seem impeccable, but there is a catch. For the environment, as Fisher conceives it, includes not just the external environment but also the 'genic environment'—*which is itself affected by natural selection.* This creates an immediate complication. For when natural selection leads gene frequencies to change, it may also modify the environment. What meaning then attaches to Fisher's idea of isolating the change in $\overline{w}$ due to natural selection itself, independently of the effect of the environment? Is such a separation really possible? And if so, what does it teach us?

3.3.1 *The FTNS explained*

The FTNS relates the partial change $\Delta\overline{w}_{NS}$ to the additive genetic variance in fitness (the modern name for Fisher's 'genetic variance'). To understand the latter notion, consider firstly the total variance in fitness in a population at a time. This is simply the average, over all n individuals, of their squared deviation from mean fitness:

$$\mathrm{Var}(w) = \frac{1}{n}\sum_i (w_i - \overline{w})^2$$

Now the fitness differences between individuals may be due to genetic or environmental factors, or both. If we know the genotype and fitness of every organism in the population, we can express Var(w) as the sum of genetic and environmental components (via a statistical technique called analysis-of-variance). Schematically, this gives:

$$\mathrm{Var}(w) = \mathrm{Var}_G(w) + \mathrm{Var}_E(w)$$

where $\mathrm{Var}_G(w)$ is the fitness variance due to genetic differences. In turn, we can divide $\mathrm{Var}_G(w)$ into additive and non-additive components:

$$\mathrm{Var}_G(w) = \mathrm{Var}_{AG}(w) + \mathrm{Var}_{NG}(w)$$

where $\text{Var}_{AG}(w)$ is the additive genetic variance. This is the fitness variation that is due to the additive, or independent, action of the genes. To understand this notion, consider how an organism's fitness depends on its genes. One possibility is that every gene has the same effect on fitness independently of which other genes the organism contains. If so, then all the genetic variance is additive, that is, $\text{Var}_G(w) = \text{Var}_{AG}(w)$. More likely, the genes will interact (in the statistical sense), that is, a gene's effect on fitness will depend partly on the genetic background. If so, there will be a non-additive component to the total genetic variance.

It is worth explaining exactly how the additive genetic variance is calculated. Consider the total set of genes (or alleles), at all loci, in the population. In a diploid species, each organism will contain 0, 1, or 2 copies of each gene. Thus we can write an organism's fitness as a linear regression on the number of copies of each gene that it contains:

$$w_i = \sum_j b_j x_{ij} + e_i$$

where w_i is the fitness of the ith organism, x_{ij} is the number of copies of the jth gene that it contains, b_j is the partial regression of fitness on the number of copies of the jth gene, and e_i is the residual. The summation is over all the alleles in the population. Essentially, this linear regression model can be thought of as an attempt to 'predict' the fitness of an organism based on information about how many copies of each gene it contains. The b_j values are chosen so as to minimize the error in prediction, that is, the sum of the squared residuals.

The quantity b_j is what Fisher called the 'average effect' of the jth gene on fitness; it measures how much difference, on average, an extra copy of the gene would make to an organism's fitness.[11] Importantly, b_j is not a fixed quantity, except in the special case where all genes act perfectly additively, but rather depends on the population's genotypic composition; so the value of b_j may change as evolution occurs.

The quantity $\sum_j b_j x_{ij}$ is what the fitness of the ith organism would be if each gene acted in a perfectly additive way; that is, if the regression model fitted perfectly. This quantity is known as the 'additive genetic value'; it may be abbreviated as AG_i, permitting the regression equation to be written as $w_i = AG_i + e_i$. In general, genes will not act perfectly additively; rather, the difference a gene makes to an organism's fitness may depend in part on the genetic background, that is, $w_i \neq AG_i$ for all i. The additive genetic variance $\text{Var}_{AG}(w)$ is then simply the variance in the AG_i values across individuals.

The partial change that Fisher is concerned with is computed as followed. In the first generation, mean fitness $\overline{w}$ is expressible as a function of the gene frequencies and the average effects. Natural selection then operates, leading gene frequencies to change; as a result $\overline{w}$ changes too, by a total of $\Delta\overline{w}$. Now $\Delta\overline{w}$ reflects partly the changed gene

[11] This is one of the two definitions of 'average effect' that Fisher uses. His other definition regresses an organism's deviation from average fitness, rather than fitness itself, on its genotypic composition. The FTNS can be proved using either definition. See Ewens and Lessard (2015) for discussion of the differences.

frequencies, and partly the fact that the average effects may themselves have changed in value; however, Fisher regards the latter change as 'environmental'. Thus $\Delta\overline{w}$ can be divided into two. The first part, $\Delta\overline{w}_{NS}$, is found from the new gene frequencies but *holding fixed the average effects at their first generation values.* The second part, $\Delta\overline{w}_{E}$, arises from the changes to the average effects themselves. Fisher's concern is solely with the first part.

The FTNS then states that the partial change $\Delta\overline{w}_{NS}$ is equal to the additive genetic variance in the first generation, divided by $\overline{w}$.[12]

$$\Delta\overline{w}_{NS} = \mathrm{Var}_{AG}(w)/\overline{w}$$

This implies that the partial change $\Delta\overline{w}_{NS}$ is always non-negative, and must be positive unless the additive genetic variance is zero. Thus natural selection has an improving tendency, Fisher argues.

3.3.2 One-locus two-allele example

A simple example can help to illustrate these concepts. Suppose that an organism's fitness depends on its genotype at a single locus, at which there are two alleles A and B. The three genotypes AA, AB, and BB have initial frequencies of P, $2Q$, and R respectively, where $P + 2Q + R = 1$. Mating is not necessarily random. The initial frequency of the A allele is $p = P + Q$ and of the B allele is $q = Q + R$.

The genotypes differ in viability; their absolute fitnesses are w_{AA}, w_{AB}, and w_{BB}, where these denote the number of surviving offspring produced by an organism of the genotype in question. Generations are non-overlapping, and genotype fitnesses are fixed. The population's average fitness is $\overline{w} = Pw_{AA} + 2Qw_{AB} + Rw_{BB}$.

The average effects of the two alleles, b_A and b_B, are found from the linear regression procedure explained above. The additive genetic values of the three genotypes are then $2b_A$, $(b_A + b_B)$, and $2b_B$; think of these as 'predicted' genotype fitnesses. The values of b_A and b_B are chosen so as to make the prediction as good as possible, that is, to minimize the distance between the actual and predicted genotype fitnesses, by the least-squares criterion.

Table 3.1. One-locus two-allele model

Genotype	Frequency	Actual fitness	'Predicted' fitness
AA	P	w_{AA}	$2b_A$
AB	$2Q$	w_{AB}	$(b_A + b_B)$
BB	R	w_{BB}	$2b_B$

[12] The division by $\overline{w}$ is necessary in the discrete-time version of the FTNS. Thus the partial change is proportional to, rather than equal to, the additive genetic variance, in discrete-time models. This does not affect the conceptual points.

If both alleles make perfectly additive contributions to fitness, that is, $(w_{AA} - w_{AB}) = (w_{AB} - w_{BB})$, then the average effects b_A and b_B can be chosen to make the actual and predicted genotype fitnesses coincide, that is, $b_A = \frac{1}{2}w_{AA}$ and $b_B = \frac{1}{2}w_{BB}$. With perfect additivity, therefore, the values of b_A and b_B depend only on the fitness parameters and not on the population's genetic composition, so will remain unchanged over generations. But in the general case this is not so.

To see why, note that the quantity $(b_A - b_B)$ is a measure of the effect on an organism's fitness, on average in the population, of substituting a *B* allele with an *A* allele—for this increases by one the organism's dosage of *A* alleles while reducing by one its dosage of *B*s. This quantity is known as the 'average effect of a gene substitution', and turns out to be a weighted average of the genotype fitness differences $(w_{AA} - w_{AB})$ and $(w_{AB} - w_{BB})$, where the weights are functions of the genotype frequencies P, $2Q$ and R (Fisher 1941). This is intuitive: for $(w_{AA} - w_{AB})$ is the effect of a $B \rightarrow A$ substitution on fitness when the substituted *B* allele is from an *AB* heterozygote, while $(w_{AB} - w_{BB})$ is the effect on fitness when the substituted *B* allele is from a *BB* homozygote. If these two quantities differ in value, the *average* effect of such a substitution will thus depend on the relative proportions of the different genotypes in the population.

Since the genotypes differ in fitness, natural selection leads to new genotype frequencies $P', 2Q'$, and R' in the second generation; so $p' = P' + Q'$ is the new frequency of the *A* allele. This leads average fitness to increase by a total of $\Delta\overline{w}$, which, following Fisher, we want to express as the sum of two parts, due to 'natural selection' and 'the environment', respectively. To find this partition, we begin with average fitness in the first generation:

$$\overline{w} = Pw_{AA} + 2Qw_{AB} + Rw_{BB} \tag{3.3}$$

We then replace each genotype fitness with its corresponding additive genetic value (or predicted fitness) in equation (3.3). This replacement necessarily leaves $\overline{w}$ unchanged, in virtue of how the average effects are defined. Therefore:

$$\begin{aligned} \overline{w} &= P2b_A + 2Q(b_A + b_B) + R2b_B \\ &= pb_A + qb_B \end{aligned} \tag{3.4}$$

In just the same way, average fitness in the second generation can be expressed in terms of the new gene frequencies and the new average effects:

$$\overline{w'} = p'b'_A + q'b'_B \tag{3.5}$$

where primes denote the second generation values. The total change $\Delta\overline{w}$ is then found by subtracting equation (3.4) from equation (3.5). With minor re-arranging, this gives:

$$\Delta\overline{w} = \overbrace{(p' - p)b_A + (q' - q)b_B}^{\Delta\overline{w}_{NS}} + \overbrace{p'(b'_A - b_A) + q'(b'_B - b_B)}^{\Delta\overline{w}_E} \tag{3.6}$$

which is the partition that Fisher is after. Note that the first term on the right-hand side, $\Delta\overline{w}_{\text{NS}}$, is found by allowing gene frequencies to change from p and q to p' and q', while holding fixed the average effects. It therefore expresses what the total change would have been had the average effects stayed constant over generations. The second term, $\Delta\overline{w}_{\text{E}}$, reflects the changes in the average effects themselves.

The partial change $\Delta\overline{w}_{\text{NS}}$ can be put in a slightly different form (by using the identities $q = 1 - p$ and $q' = 1 - p'$):

$$\Delta\overline{w}_{\text{NS}} = 2(p' - p)(b_A - b_B) \tag{3.7}$$

This is the form given by Fisher (1941), and is highly intuitive. For $(b_A - b_B)$, as we have seen, is the average effect on fitness of a $B \rightarrow A$ substitution, given the population's genotypic makeup. When natural selection leads p to change to p', this corresponds to making $2(p' - p)$ allelic substitutions per capita (since each organism is diploid). So the product of $2(p' - p)$ and $(b_A - b_B)$ equals the partial change $\Delta\overline{w}_{\text{NS}}$.

To establish the FTNS, we find the additive genetic variance in fitness (in the first generation). This is the variance of the additive genetic values:

$$\text{Var}_{\text{AG}} = P(2b_A)^2 + 2Q(b_A + b_B)^2 + R(2b_B)^2 - \overline{w}^2 \tag{3.8}$$

Straightforward algebra then shows that the partial change in equation (3.7) equals the additive genetic variance in equation (3.8) divided by $\overline{w}$, which proves the FTNS for this simple case (Appendix 3.1).

3.3.3 Environment and causality

The FTNS is straightforward mathematically, but its significance (if any) is less clear. Does the partial change $\Delta\overline{w}_{\text{NS}}$ have a natural biological meaning, and is it a quantity we should care about? There are two related issues here. The first concerns Fisher's device of holding fixed the average effects as gene frequencies change. The second concerns whether $\Delta\overline{w}_{\text{NS}}$ can really be regarded as the change 'due to' natural selection. I shall address the issues in turn.

Fisher's rationale for holding fixed the average effects was to consider the impact of natural selection on $\overline{w}$ in a 'constant environment'. In itself this sounds perfectly reasonable, an instance of the familiar point that to assess the impact of one factor on an outcome of interest, all other factors must be held fixed. Since $\overline{w}$ may be affected by factors other than natural selection, surely it makes sense to hold them fixed so far as possible? However, Fisher conceives of the environment in a very particular way, such that any change in the genes' average effects implies that the environment has changed. This raises the suspicion that his 'environmental change' is a catch-all, and that the partial change $\Delta\overline{w}_{\text{NS}}$ thus lacks biological significance (cf. Price 1972b).

To focus the issue, consider why the average effects may change from one generation to another. There are two broad explanations. First, genotype fitnesses themselves may have changed, for any of a number of reasons. These include changes to the physical environment (e.g. weather), changes to other species (e.g. competitors), and frequency or density-dependence (e.g. a genotype may be advantageous when rare but not when common). Second, even if genotype fitnesses are constant, then the average effects may still change, due to non-additivity in the genetic system. For example, in the one-locus model above, if $(w_{AA} - w_{AB}) \neq (w_{AB} - w_{BB})$ then the average effects b_A and b_B will be functions of genotype frequencies, so will change as the population evolves. Empirically such non-additivities, within and between loci, are common.

The first of these factors—changing genotypic fitnesses—can certainly be viewed as environmental. That a genotype's fitness depends on the environment is a commonplace in biology, as is the idea that the environment includes the population's genotypic composition. This notion of environment is standard where organisms engage in fitness-affecting interactions with one another. However, the second factor—non-additive genetics—is a different matter. Treating changes in the average effects that arise for this reason as environmental is dubious, as critics have noted (Price 1972b). For it commits Fisher to saying that even in the *absence* of social interactions, that is, if an individual's fitness depends solely on its own genotype, nonetheless the selective environment changes as the population's composition evolves. This is an odd notion, and seemingly unmotivated.

Can anything be said in Fisher's defence? I think that it can. I suggest we regard Fisher's argument as a precursor of the gene's eye view of evolution later articulated by Williams (1966) and Dawkins (1976).[13] On this view, natural selection is seen as choosing between genes which happen to be temporarily packaged into pairs in diploid organisms. From a gene's point of view, how the packaging into pairs is done, which depends on the mating pattern, is part of the social environment. Now if there are non-additive effects on genotype fitness, then how the genes get packaged into pairs will influence the average effects. Changes to the average effects that derive from non-additivity thus reflect changes to the social environment of the genes (Okasha 2008).

To appreciate this argument, consider again the one-locus model above. Note that the partial change $\Delta\overline{w}_{\rm NS}$ in equation (3.6) is computed from the change in *gene* frequencies (and the first generation average effects); we do not need to know the change in *genotype* frequencies. That is, how the genes get packaged into organisms in the second generation is irrelevant to the partial change $\Delta\overline{w}_{\rm NS}$. However, this packaging does affect the total change, since the average effects in the second generation, and thus the value of $\Delta\overline{w}_{\rm E}$, depend on the new genotype frequencies, not merely the gene frequencies. Thus the manner in which the genes are packaged into organisms feeds into the second, environmental component of the total change $\Delta\overline{w}$.

There is some evidence that Fisher thought in these terms, even though he did not lay out the gene's eye view explicitly. He certainly regarded the mating pattern as an environmental factor, and explicitly refers to a 'constant genic environment'

[13] This suggestion is developed in Okasha (2008) and Edwards (2014).

in describing the partial change in mean fitness that he was concerned with (1941, p. 56). What Fisher meant by this phrase was actually quite subtle. One *might* think that a constant genic environment, in the one-locus model, means that the A allele is found in heterozygotic and homozygotic environments in unchanging proportions, that is, $P/Q = P'/Q'$; and similarly for the B allele. If so, this would mean that the genic environment of at least one of the alleles necessarily changes whenever the genotype frequencies change.[14] This is how the notion of 'genic environment' is often understood in discussions of the gene's eye view.[15]

However, Fisher's notion was different. In the one-locus two-allele model, Fisher (1941) introduces a quantity called λ which is defined as:

$$\lambda = Q^2/PR$$

That is, λ is the ratio of the square of half the heterozygote frequency to the product of the homozygote frequencies. Note that if mating is random, and thus genotype frequencies are in Hardy-Weinberg proportions, then $\lambda = 1$. If mating is assortative then $\lambda > 1$, as the proportion of heterozygotes will be reduced. Therefore λ is a measure of the deviation from random mating in the population, in a given generation.

Fisher then defines constancy of the genic environment as constancy of λ. The rationale behind this definition is simply the fact that if λ stays constant across generations then the average effects will stay constant too, and the partial change $\Delta\overline{w}_{NS}$ will equal the total change $\Delta\overline{w}$; whereas if λ changes across generations, then except in special cases (such as perfect additivity), the partial and total changes will differ. So the 'genic environment', as understood by Fisher, does not necessarily change whenever the population evolves; it all depends on what happens to λ.

The upshot, therefore, is that Fisher's notion of environmental constancy, and his device of holding fixed the average effects as gene frequencies change, is not devoid of biological meaning, as some critics suggest. Rather, it is an expression of a gene's eye view of evolution. For holding fixed the average effects equates to holding λ fixed, which in turn equates to keeping the mating pattern the same—and the latter is part of a gene's social environment. Of course, the gene's eye view has its critics; my point is not to defend it, but rather to show that it offers a plausible way of defending the FTNS from the objection that it relies on an unmotivated concept of environment.

On the other hand, Fisher offers no independent reason for why the population's mating pattern should be quantified by λ. There exist other, non-equivalent measures of deviation from random mating in population genetics (such as the 'coefficient of inbreeding'), and Fisher offers no justification for choosing λ as the preferred measure.[16] Rather, he simply shows that constancy of λ is implied if we equate the

[14] The point here is that it is impossible that both $P/Q = P'/Q'$ and $Q/R = Q'/R'$, given that $P + 2Q + R = P' + 2Q' + R' = 1$.

[15] See for example Sterelny and Kitcher (1988).

[16] The coefficient of inbreeding is originally due to Wright (1922), and is widely used. See for example Rice (2004) pp. 104–7.

partial change $\Delta\overline{w}_{NS}$ with the total change $\Delta\overline{w}$. Thus although Fisher's procedure is biologically meaningful, it does still contain an element of arbitrariness; for the particular mathematical definition of 'constant genic environment' on which it relies is not the only possible one, and has been chosen to yield the desired result. Therefore, even if we accept a gene's eye view of evolution, this does not fully justify Fisher's conception.

Turning to the second issue, can $\Delta\overline{w}_{NS}$ be regarded as the change 'due to' or 'caused by' natural selection, as Fisher intimated? This causal interpretation, which many commentators endorse (e.g. Frank and Slatkin 1992), seems essential if the FTNS is be to taken as showing that natural selection has an optimizing tendency. But is it valid?

Ewens (1989, 2010) explicitly rejects the identification of $\Delta\overline{w}_{NS}$ with the change caused by natural selection. His reason is simple. $\Delta\overline{w}_{NS}$ is computed by holding fixed the average effects; however, the average effects depend on the genotype frequencies, which themselves are affected by natural selection. Thus if genotype frequencies change solely because of natural selection (as opposed to factors such as drift and migration), and if this leads the average effects to change, then the total change $\Delta\overline{w}$, not the partial change $\Delta\overline{w}_{NS}$, is attributable to natural selection, by Ewens' lights. Thus the causal construal of the FTNS is untenable, he argues.

This argument is powerful but not conclusive. For there is a possible response that a defender of Fisher can make, by combining his notion of genic environment with a distinction between the direct and indirect effects of natural selection. The idea here is that when selection occurs, leading gene frequencies to change, this directly affects $\overline{w}$; but it may also change the environment, thus indirectly affecting $\overline{w}$ via a second route. Thus there are two causal pathways from natural selection to $\overline{w}$ (Figure 3.3). The partial change $\Delta\overline{w}_{NS}$ is the change transmitted along the direct route only, and is to be found by blocking the indirect route, that is, holding fixed the environment. Therefore $\Delta\overline{w}_{NS}$ is the change *directly* caused by natural selection.

How convincing is this response? The general point that one variable may transmit causal influence to another via multiple pathways is uncontroversial. Adding fertilizer to a field of corn plants may increase the yield by improving soil quality, and also by poisoning the rats that eat the plants. It is also uncontroversial that to measure the causal influence along one path we need to block the other pathways; this may be

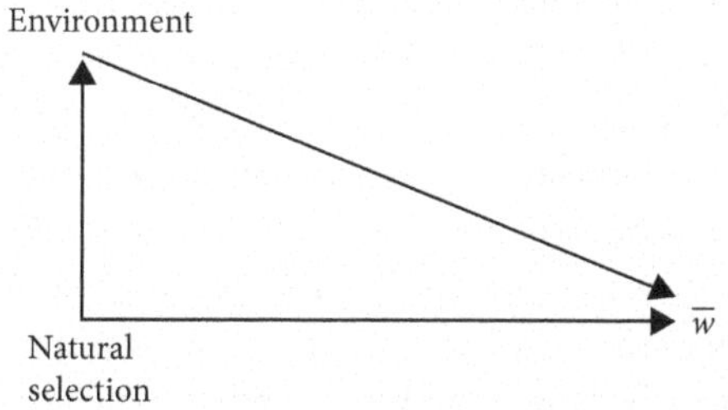

Figure 3.3. Two causal pathways

done either experimentally or statistically (i.e. by holding fixed the relevant variables). Moreover, the idea that natural selection may have indirect effects on an evolving population, via changing its environment, is familiar in other areas of biological discussion, for example, of 'niche construction' (Odling-Smee *et al.* 2003). So the direct/indirect distinction, and the technique for measuring the direct effect, are unexceptionable.

The two causal pathways idea thus provides the ingredients to salvage the causal interpretation of the FTNS from Ewens' critique. However, this presumes that the causal graph in Figure 3.3 is a correct depiction of the causal forces actually at work on a population, which requires that the environmental variable is real and does actually influence $\overline{w}$. So ultimately, the issue reduces to whether we accept Fisher's broad notion of environment, including the specific definition of constant genic environment that he relies on (constant λ). Now the former is defensible from a gene's eye viewpoint, as we have seen; however, the latter lacks a fully compelling justification. Therefore the case for the causal construal of $\Delta\overline{w}_{NS}$ is not logically watertight.

Finally, even if we agree that $\Delta\overline{w}_{NS}$ is the component of change 'directly' due to natural selection, one may still wonder why this quantity matters. Given that indirect effects are likely to be ubiquitous, surely it makes more sense to care about the total change $\Delta\overline{w}$? This leads to the question of what the FTNS teaches us.

3.3.4 *What does the FTNS show?*

Fisher thought that the FTNS was an important result, comparable in some respects to the second law of thermodynamics (1930, p. 37). He took it to show that natural selection has an inherent optimizing tendency, using the expressions 'rate of progress' and 'rate of improvement' to describe the partial increase in mean fitness. Moreover, he suggests that as a result of this tendency towards 'improvement', an evolved organism will appear 'highly adapted to its place in nature' (p. 46).

Some commentators share this assessment. Thus Grafen (2003) argues that the FTNS 'isolated the adaptive engine in evolution' (p. 327). However, others dissent. For example, Price (1972b) takes the FTNS to show that natural selection 'at all times acts to increase the fitness of a species to live under the conditions that existed an instant earlier' (p. 131). He continues 'but since the standard of "fitness" changes from instant to instant, this constant improving tendency does not necessarily get anywhere as measured by any fixed standard' (p. 131). In effect, Price is saying that since the FTNS is silent about the total change $\Delta\overline{w}$, it tells us nothing about adaptation.

How should this be adjudicated? There are two issues here: what the FTNS tells us about selection (the process) and what it tell us about adaptation (the product). Regarding the former, if we accept the causal construal of $\Delta\overline{w}_{NS}$, above, then the FTNS does suggest that selection involves optimization, in the sense that its direct effect is always to increase $\overline{w}$. However, this is a rather weak sense. For nothing follows about how often the tendency for $\overline{w}$ to increase will actually be realized; and what prevents it being realized will often be natural selection itself (via its effect on the environment). So the FTNS does not vindicate the naive notion that natural selection leads inexorably to improvement, or to maximization of some quantity, for essentially the same reason that it does not vindicate the hill-climbing view of evolution. At most, it shows that selection will always tend to push a population up the adaptive landscape, but may simultaneously cause the landscape itself to move.

Recall our previous discussion of agential thinking (type 2), in which the process of natural selection is analogized to a process of deliberate choice, or metaphorically personified. We argued that vestiges of this mode of thinking still exist in biology, since the conception of natural selection on which it rests, as a goal-directed process that tends towards greater fitness, is still prevalent. At first blush the FTNS may seem to support this conception, and it is sometimes invoked in this context; however, our analysis suggests otherwise. The weak sense in which selection is an optimizing process, manifest in the partial change $\Delta\overline{w}_{NS}$ being non-negative, does not imply that Darwinian evolution is a directional or goal-directed process, nor justify describing it in teleological terms.

What about adaptation? Does the FTNS support Darwin's argument that natural selection will lead organisms to become well-adapted to the environment? Again, only weakly. For consider the many factors that can prevent, or limit, the evolution of organismic adaptation: stochasticity, changing physical environment, non-selective forces, genetic and developmental constraints. The FTNS is either silent about these factors, or lumps them together as part of 'environmental change'. At most, it shows that *if* the environment is constant, and thus if the factors that limit or prevent adaptation from evolving are absent, then organisms will evolve traits that maximize population growth rate, and hence best equip them to survive and reproduce in that environment. However, the FTNS itself implies that the antecedent of this conditional statement will often be untrue.

Recall agential thinking (type 1), in which an evolved organism is treated as akin to an agent with a goal, namely maximizing its fitness. Grafen (2003) reads Fisher as an advocate of this view, arguing that he used the FTNS as a license for 'regulated anthropomorphism' (p. 326). How committed Fisher really was to this form of agential thinking is debatable; however, in any case the FTNS does not justify it, if my analysis is correct. For agential thinking (type 1) derives its rationale from the empirical fact that organisms are often well-adapted and exhibit a unity-of-purpose; and the FTNS does not show that this is to be expected. At most, it shows that where high degrees of adaptation are found in nature, the twin hypotheses of natural selection plus environmental constancy, in a very specific sense, constitute one possible explanation.

To conclude, the FTNS does not show that organismic adaptation is an inevitable or even likely outcome of evolution by natural selection. I do not suggest that Fisher thought otherwise (though his writings are ambiguous on this point), nor that this is a lacuna that needs filling. Perhaps the FTNS is the most general, precise statement that can be made about how natural selection works. If so, then the grounds for being an adaptationist must ultimately be empirical, not theoretical. This point is revisited in section 4.4 in the next chapter.

3.4 Conclusion

The idea that natural selection has an optimizing tendency is *prima facie* compelling, and can be traced to Darwin himself. It fits naturally with the popular notion that evolution by natural selection is a directional process, tending toward 'improvement', adaptation, or higher fitness. We have examined two influential attempts to make this notion precise, by Wright and Fisher, but found that neither is entirely successful. The hill-climbing property is not a universal feature of Darwinian evolution; and even when it does obtain, the steepest-ascent property usually does not. The FTNS points only to a weak sense in which natural selection is optimizing, and requires that 'environmental constancy' be understand in a specific way. So although the adaptive landscape and the FTNS are both useful concepts, neither provides a general vindication of the proposition that natural selection will tend to produce adaptation, even in the best-case scenario of frequency-independent selection.

This has implications for both of the types of agential thinking that we discussed in chapter 1. The fact that natural selection does not in general maximize mean population fitness, nor 'tend' to do so, undermines the view of Darwinian evolution as goal-directed; this tells against agential thinking (type 2), the metaphorical identification of natural selection with a process of deliberate choice. In itself, this leaves untouched agential thinking (type 1), the practice of treating an evolved organism as akin to an agent with a goal, for this is a way of thinking about adaptation, not selection. However, it does suggest that the justification for this mode of thinking does not derive from the logic of natural selection itself.

Appendix 3.1

Proof of the FTNS for the one-locus model of section 3.1.1

Initial frequency of A allele: $p = P + Q$
Initial frequency of B allele: $q = Q + R$

Marginal fitnesses of A and B alleles:

$$\begin{aligned} w_A &= (Pw_{AA} + Qw_{AB})/(P+Q) \\ w_B &= (Rw_{BB} + Qw_{AB})/(R+Q) \end{aligned} \tag{A.1}$$

Average effects b_A and b_B are chosen so that:

$$P(2b_A - w_{AA}) + Q(b_A + b_B - w_{AB}) = 0 \tag{A.2}$$

and

$$R(2b_B - w_{BB}) + Q(b_A + b_B - w_{AB}) = 0 \tag{A.3}$$

Expanding equation (A.2) and re-arranging gives:

$$\begin{aligned} &P2b_A + Qb_A + Qb_B = Pw_{AA} + Qw_{AB} \\ \Longrightarrow\ &P2b_A + Qb_A + Qb_B = pw_A \\ \Longrightarrow\ &pb_A + (Pb_A + Qb_B) = pw_A \\ \Longrightarrow\ &Pb_A + Qb_B = p(w_A - b_A) \\ \Longrightarrow\ &P(b_A)^2 + Qb_Ab_B = p(w_A - b_A)b_A \end{aligned} \tag{A.4}$$

Similarly, expanding and re-arranging equation (A.3) gives:

$$R(b_B)^2 + Qb_Ab_B = q(w_B - b_B)b_B \tag{A.5}$$

Adding equations (A.4) and (A.5):

$$P(b_A)^2 + 2Qb_Ab_B + R(b_B)^2 = p(w_A - b_A)b_A + q(w_B - b_B)b_B \tag{A.6}$$

Average fitness in first generation:

$$\begin{aligned} \overline{w} &= Pw_{AA} + 2Qw_{AB} + Rw_{BB} \\ &= 2b_Ap + 2b_Bq \end{aligned} \tag{A.7}$$

Partial change in average fitness:

$$\begin{aligned} \Delta\overline{w}_{NS} &= (P' - P)2b_A + 2(Q' - Q)(b_A + b_B) + (R' - R)2b_B \\ &= 2(p' - p)(b_A - b_B) \\ &= 2b_Ap' + 2b_Bq' - \overline{w} \qquad \text{(from A.7)} \end{aligned} \tag{A.8}$$

Additive genetic variance in fitness (in the first generation):

$$\begin{aligned} \mathrm{Var}_{AG} &= P(2b_A)^2 + 2Q(b_A + b_B)^2 + R(2b_B)^2 - \overline{w}^2 \\ &= 2p(b_A)^2 + 2q(b_B)^2 + 2(P(b_A)^2 + 2Q(b_Ab_B) + R(b_B)^2) - \overline{w}^2 \\ &= 2p(b_A)^2 + 2q(b_B)^2 + 2(p(w_A - b_A)b_A + q(w_B - b_B)b_B) - \overline{w}^2 \\ &= 2pb_Aw_A + 2qb_Bw_B - \overline{w}^2 \qquad \text{(from A.6)} \\ &= 2b_Ap'\overline{w} + 2b_Bq'\overline{w} - \overline{w}^2 \quad \left(\text{since } p' = p\frac{w_A}{\overline{w}};\quad q' = q\frac{w_B}{\overline{w}}\right) \end{aligned} \tag{A.9}$$

From equations (A.8) and (A.9):

$$\Delta\overline{w}_{NS} = \mathrm{Var}_{AG}/\overline{w}$$

4

Grafen's Formal Darwinism, Adaptive Dynamics

4.1 Introduction

The previous chapter examined two classical treatments of fitness maximization in evolutionary biology: Wright's adaptive landscape and Fisher's FTNS. We argued that both could be seen as formal expressions of Darwin's argument that natural selection will lead to well-adapted individuals and thus has an optimizing tendency. However, they are rather indirect expressions of Darwin's argument, since their focus is on the change and partial change, respectively, in a population-level property, namely mean fitness. We argued that this does not automatically disqualify them from teaching us something about individual-level adaptation, given that a population's mean fitness depends on the fitness of its constituent individuals. But a more direct expression of Darwin's argument would focus on the effect of natural selection on individuals themselves, rather than on an aggregate population property.

Such an argument has recently been given by Alan Grafen as part of his ongoing 'formal Darwinism' project.[1] Grafen's stated aim is to vindicate the assumption of individual fitness maximization that he sees at work in the day-to-day practice of evolutionary biologists, especially in behavioural ecology. To this end, he argues that an 'individual-as-maximizing-agent analogy' can be derived from evolutionary first principles. Grafen (2014a) describes this as a formalization of Darwin's argument that natural selection will lead to the appearance of design in nature, and a justification of something that many biologists take for granted.

Grafen's project has philosophical as well as biological significance, for it relates directly to agential thinking (type 1), the practice of treating evolved organisms as akin to agents trying to achieve a goal or maximize a utility function, and making them the subjects of intentional attributions. In chapter 1 we argued that agential thinking (type 1) is valid to the extent that evolved organisms are in fact well-adapted and exhibit a unity-of-purpose. By contrast, Grafen aims to supply a theoretical rather than an empirical justification for agential thinking, by deducing the proposition that organisms will be agent-like, or well-adapted, from evolutionary first

[1] See in particular Grafen (2002, 2006, 2008, 2014a).

principles. However, Grafen's project is controversial. A special issue of the journal *Biology and Philosophy* in 2014 contains divergent assessments of the project by leading evolutionists; some regard it as a proven success, others as wrong-headed.[2] In section 4.2, I lay out the structure of Grafen's argument and offer my own assessment of what it shows.

The focus in section 4.3 shifts to frequency-dependent selection. Grafen's ideas, like those of Wright and Fisher, pertain mainly to frequency-independent selection, which is the traditional home for the doctrine of fitness-maximization (though in principle the FTNS applies more broadly). However, since frequency-dependent selection is quite common in nature, it is important to ask whether an argument similar to Grafen's could be extended to this case. Section 4.3 looks briefly at this question by drawing on work in the field of adaptive dynamics, which is an abstract framework for analysing frequency-dependent selection. Section 4.4 steps back and extracts a general moral concerning the status of adaptationism in biology, and the validity of agential thinking. Section 4.5 concludes.

4.2 Grafen's 'Maximizing Agent' Analogy

Grafen's formal Darwinism project offers a novel way of expressing, and trying to justify, the idea that natural selection will lead to well-adapted individuals. His starting point is the notion of an 'optimization programme', which he uses to capture the Darwinian idea that an evolved individual will have phenotypic traits that are fitness-maximizing, or further their goal of survival and reproduction better than the alternatives.

Grafen's optimization program consists of a real-valued 'objective function' that the individual tries to maximize by means of wielding an 'instrument', namely its phenotype. The domain of the objective function is a set of phenotypes. If the individual has a phenotype that maximizes the objective function they are said to be optimal, or to 'solve the optimization programme'. The objective function is meant to represent the individual's biological fitness, but Grafen deliberately leaves open how this is defined; the appropriate definition is something he wants to derive. The intended interpretation is that the objective function represents the 'goal' of individual behaviour, in the sense that if individuals are well-adapted, they will behave as if they are trying to maximize the function. Thus there is a close conceptual link between Grafen's optimization programme and the rational choice formalism developed in section 1.7 to capture agential thinking (type 1).

Importantly, the 'optimal' phenotype in Grafen's analysis is relative to a given set, which is taken as given. (The intended interpretation is that it contains all 'biologically feasible' phenotypes.) This is not in itself a problem, and is necessary for Grafen's

[2] *Biology and Philosophy* volume 29, 2014.

argument to have the desired generality. It means that when Grafen defines adaptation in terms of optimality, he is not necessarily at odds with traditional opponents of the idea that evolved organisms will be optimal, for they often take 'optimal' to mean better than any *conceivable* alternative. The fact that organisms are rarely if ever optimal in this strong sense is compatible with their being well-adapted, which is what Grafen is trying to express.

Grafen's aim is to connect the process of natural selection with the optimality, in his sense, of the individuals in the population. To this end, he constructs a series of population-genetic models of evolution. In the simplest model, generations are non-overlapping and an individual's reproductive success depends only on its phenotype, which in turn depends on its genotype—thus social interaction is absent, as is frequency-dependence (Grafen 2002, 2014a). Grafen then uses the Price equation to express the change in frequency of any gene in the population over one generation. This gives:

$$\Delta p = \frac{1}{\overline{w}} \left\{ \mathrm{Cov}(w_i, p_i) + \mathrm{Exp}(w_i \delta_i) \right\} \tag{4.1}$$

where w_i is the number of successful gametes of the ith individual; p_i is the frequency of the gene at its locus within the ith individual (which equals 0, $\frac{1}{2}$ or 1 for diploids); δ_i is the difference in gene frequency between the ith individual and the average for its offspring; p is the population-wide frequency of the gene; and $\overline{w}$ is the average gametic contribution, or fitness, in the population (Price 1970). The covariance and expectation are taken over all individuals in the population.

On the usual interpretation of the Price equation, the two terms on the right-hand side correspond to the effects of natural selection and 'transmission bias' on the total evolutionary change, respectively (Frank 1998, Okasha 2006). In the case at hand, where selection is acting on individuals, transmission bias refers to changes in gene frequency arising during reproduction, because of either mutation or gametic selection (meiotic drive). In the simplest version of his argument, Grafen assumes that mutation and meiotic drive are absent, enabling him to drop the second 'Exp' term.[3] This means that, by assumption, differential survival or fecundity of individuals is the only cause of gene frequency change.

Grafen next defines two technical terms. There is 'scope for selection' in a population if there is some gene in the population that will change in frequency from one generation to the next, that is, $\Delta p \neq 0$ for all genes p. There is 'potential for selection' if a mutant gene, coding dominantly for one of the possible phenotypes, would be able to spread, that is, the population is invasible. (The point of the qualification 'dominantly' will become clear.) Note that potential for selection, so defined, refers to what *would* happen to a mutant it if did arise; thus the presence of potential for selection is quite compatible with the transmission bias in the population, at any time, being zero. If there is neither scope nor potential for selection, we may say that the

[3] Strictly speaking, it is the expected transmission bias that will be zero.

population is at selective equilibrium (my terminology). Note that equilibrium in this sense, that is, zero gene frequency change plus uninvasibility, does not imply that the population will necessarily evolve towards the equilibrium from an arbitrary starting point.

The key feature of Grafen's set-up is that selection and optimality have been defined separately, using different concepts—the latter in terms of an individual's objective function, the former in terms of gene frequency change via the Price equation. So it is an open question how the two relate. Grafen formulates a number of 'selection-optimality' links, which are 'if . . . then' statements linking the two concepts (Box 4.1).[4] The links thus posit logical connections between the process of natural selection, described in terms of gene frequency change, and the phenomenon of phenotypic adaptation, that is, individuals having optimal phenotypes. Grafen's aim is then to find a choice of objective function which makes his links come out true.

Box 4.1. Grafen's selection-optimality links

1. If all individuals in the population are optimal, then selective equilibrium obtains.
2. If all individuals are equally sub-optimal, there is no scope for selection but there is potential for selection.
3. If individuals vary in their optimality, the change in the frequency of any gene equals its covariance with the individuals' attained values of the objective function.
4. If selective equilibrium obtains, then all individuals in the population are optimal.

Source: based on Grafen (2014a), p. 164

Intuitively, it is easy to see why links 1 and 2 will come out true, if the objective function is suitably defined in terms of some measure of biological fitness. Link 1 says that if all individuals in the population optimize the objective function, then selective equilibrium obtains. Now clearly, if all individuals in a population have a phenotype that confers maximum fitness, there will be no variation for selection to act on, hence no scope for selection, and no mutants will be able to invade, hence no potential for selection. Link 2 says that if all individuals are equally sub-optimal, there is no scope for selection but there is potential for selection. Again, if all individuals attain the same fitness value, there will be no fitness variation hence no scope for selection; but if that value is less than the maximum possible, a mutant coding dominantly for a superior phenotype will be able to invade; hence there is potential for selection. Link 3 says

[4] Grafen's links are formulated slightly differently in different papers; see Okasha and Paternotte (2012, 2014) for discussion of the differences.

that if individuals vary in optimality, then the change in frequency of any gene equals its covariance with the individuals' attained values of the objective function. This is less intuitive, but inspection of the Price equation (4.1), with the second 'Exp' term set to zero, shows why it will be true, if the objective function is defined in terms of individual fitness w_i. Link 4, which is the converse of link 1, is rather unintuitive, and indeed might be thought to conflict with the point, familiar from basic population genetics, that natural selection will not always eliminate sub-optimal individuals from a population. How then can it be true that selective equilibrium implies individual optimality, as link 4 says? The answer turns on Grafen's definition of equilibrium, as section 4.2.2 explains.

The novelty of Grafen's project lies in his suggestion that links 1–4 capture Darwin's argument that natural selection will give rise to well-adapted individuals, whose phenotypes maximize their fitness, and in the idea that 'fitness' should be defined as whatever the objective function needs to be, to secure the links' truth. Importantly, the requisite definition will depend on the evolutionary model. Thus consider three of Grafen's main results. First, in the simple model without social interaction described earlier, if the individual's objective function is defined as $(w_i/\overline{w})$, that is, number of successful gametes divided by the population average, or relative fitness, then the links will hold (Grafen 2002). Second, in a model with (additive) social interactions, the individual's objective function needs to be defined as relative 'inclusive fitness' in the sense of Hamilton (1964), for the links to hold (Grafen 2006). Third, in a simple model with uncertainty, in which an individual's fitness is a random variable, the objective function needs to be defined as *expected* relative fitness (Grafen 1999). The first of these results is fairly obvious, at least for links 1–3; the last two are not obvious and moreover rely essentially on particular model assumptions, as we shall see when we examine social evolution and uncertainty in chapters 5 and 8, respectively.

4.2.1 Four comments on Grafen

Grafen's project has many merits, four of which deserve mention. First, it highlights the logical difference between the process of evolution by natural selection, which is the concern of population genetics and evolutionary dynamics, and the phenomenon of phenotypic adaptation, which is the concern of evolutionary biologists in the field. This logical gap is not always recognized (in part because of the tendency to define an adaptation as 'a trait that evolves by natural selection'); asking how it can be bridged is an important project. In particular, Grafen's approach is a useful corrective to the tendency, both within evolutionary biology and beyond, simply to *assume* that natural selection will lead to phenotypic adaptation, at least if unchecked by opposing evolutionary forces; however, this is a substantive proposition, not something that should be taken for granted.

Second, Grafen's analysis highlights the fact that we cannot start from an a priori assumption about what an individual's 'goal' is, and then hope to explain all its evolved traits as contributions to that goal. If one assumes that the individual's goal is maximizing its personal reproductive success, for example, then altruistic traits will not be readily explained as contributing to the goal; so agential thinking requires the appropriate choice of goal. However, this raises the question of what 'appropriate' actually means. Requiring that the individual's objective function be chosen so that the four selection-optimality links come out true represents one possible answer to this question. In effect, this is to suggest that we may regard Grafen's links as axiomatic conditions that the concept of individual fitness should satisfy.

This prompts a further question: can we regard Grafen's selection-optimality links as fully characterizing, that is, implicitly defining, biological fitness? If so this would be a result of some interest, helping to resolve the thorny question of how exactly fitness should be defined in different biological scenarios, and why. The answer depends on whether a *single* objective function will satisfy links 1–4, in any given biological model. Grafen himself expresses uncertainty on this point, arguing that the uniqueness of the objective function is a 'tentative' conjecture (2014a, p. 12). However, in fact it follows from link 3 alone that the objective function is very nearly unique. For if we suppose that two objective functions f and g both satisfy link 3, it follows from the Price equation (4.1) (with the 'Exp' term set to zero) that $b_{fp} = b_{gp}$, that is, the linear regression of f_i on p_i must be equal to that of g_i on p_i, where f_i and g_i are the values of f and g for the ith individual in the population, and p_i is their individual gene frequency.[5] Now the only way that this equality can hold for *every* gene p, irrespective of its pattern of association with f and g, is if $b_{fg} = b_{gf} = 1$, which in turn implies that $f = g + c$, that is, the objective function is unique up to addition of a constant. So it is indeed plausible to regard Grafen's links as implicitly defining fitness.

Third, though Grafen's main concern is with individual organisms as the maximizing agents, his methodology can in principle be applied to other entities, for example, groups or genes.[6] Thus the project offers a quite general way of trying to justify a form of agential thinking (type 1) in biology, which involves no essential commitment to the identity of the agent nor to the goal that they are pursuing. The application to groups was briefly discussed in section 2.3.

Fourth, Grafen's analysis dovetails neatly with our discussion of agential thinking in chapter 1, in particular with our unity-of-purpose constraint on agency. To recall, this constraint says that an individual organism can only be treated as agent-like, for the purposes of evolutionary analysis, to the extent that its evolved traits have complementary rather than antagonistic functions. We argued for this by exploring the presuppositions of the notion of agency, and illustrating how agential thinking can lead astray if an organism does not exhibit unity-of-purpose, as for example if it is riven by intra-organismic conflict. Interestingly, the unity-of-purpose constraint emerges as a logical consequence of Grafen's analysis, in the following precise sense. Unless the empirical preconditions for unity-of-purpose are satisfied, such as the

[5] To see this, note that by the definition of covariance, $\text{Cov}(f, p) = b_{fp} \cdot \text{Var}(p)$.

[6] Gardner and Grafen (2009) apply the methodology to groups, and Gardner and Welch (2011) to genes.

absence of within-individual genetic conflict, then Grafen's selection-optimality links cannot all hold true. So if those links do hold true, then the unity-of-purpose constraint will automatically be satisfied. This point is worth explaining.

The key point is that in order for all of the selection-optimality links to hold true, the second 'Exp' term of the Price equation (4.1) must be zero for every gene p. This is easily seen in the simple non-social model described previously. For suppose the 'Exp' term is non-zero for some gene p. Then, even if all individuals produce the same number of successful gametes, that is, $w_i = w_j$ for all individuals i, j (which implies that the 'Cov' term is zero), Δp will be non-zero, so there is 'scope for selection'. Therefore link 1 is false, for optimality of all the individuals does not imply that selective equilibrium obtains. Thus the truth of link 1 requires that the 'Exp' term be zero for every gene; this in turn implies that meiotic drive, and thus within-individual genetic conflict, must be entirely absent. Therefore if Grafen's individual-as-maximizing-agent analogy obtains, that is, if all of his selection-optimality links hold true, then individuals' genes will necessarily encode phenotypic traits that conduce to a single goal and thus have complementary rather than antagonistic functions.

It is quite striking that while we defended the unity-of-purpose constraint in section 1.7 on philosophical grounds, by reflecting on the notion of agency and what it presupposes, Grafen's analysis yields the constraint as a deductive consequence, in the sense that his selection-optimality links will only hold true if within-individual selection, and thus intra-genomic conflict, are absent. I take this both as a point in favour of our unity-of-purpose constraint on agency, and as suggestive evidence that Grafen's analysis is onto something important.

4.2.2 What do Grafen's results show?

Grafen's results raise two issues of interpretation. First, how do they square with other results in evolutionary theory which suggest that fitness-maximization is not a universal outcome of evolution by natural selection, but only obtains under restrictive conditions? Second, do his results vindicate the adaptationist assumptions of (some) working biologists, and constitute a formal expression of Darwin's argument in the *Origin*, as he claims? I look at these in turn.

Grafen (2014a) takes his selection-optimality links to show that there is 'a very general expectation of something close to fitness maximization' in evolution (p. 166). This is a striking claim, since it appears to contradict a claim made by many other evolutionists, who oppose the idea that natural selection will necessarily lead to optimization, or fitness-maximization.[7] What explains this divergence?

I think the answer is four-fold. First, Grafen's concern is with individual rather than mean fitness maximization, which as we have seen in chapter 3 are different concepts. So the fact that selection does not necessarily drive a population up a

[7] See for example Moran (1964), Rice (2004), and Metz *et al.* (2008).

mean fitness gradient, which is the basis for some of the opposition to optimization, does not in itself undermine Grafen's project.[8] Second, genetic constraints are one reason why optimal phenotypes may not evolve; however, it turns out that Grafen's selection-optimality links hold true even in the presence of genetic constraints, as we will discuss in a moment. Third, Grafen's selection-optimality links connect selective equilibrium with individuals' having the optimal phenotype; but the links do not imply that the equilibrium will actually be attained, even if natural selection is the only evolutionary force at work. However, opponents of fitness-maximization have precisely been concerned to point out that selection need not lead to a stable equilibrium, for example, because gene frequencies may cycle indefinitely. Grafen's formulation of the question bypasses this concern.

Fourth, and perhaps most importantly, Grafen is concerned with frequency-independent selection, while opponents of optimization often focus on frequency-dependence. (This is why the inclusive fitness results of Grafen (2006) require that all social interactions are additive; see section 5.4.) To see this, note that Grafen's objective function takes a single argument, namely the individual's phenotype. This means that game-theoretic scenarios, where the relevant notion of optimality is having the optimal phenotype conditional on the phenotypes of others, are not covered. A population undergoing frequency-dependent selection need not reach equilibrium at all; and if it does, the equilibrium cannot generally be characterized in terms of individuals maximizing a function whose sole argument is their own phenotype. This point is treated in detail in section 4.3.

These four points explain how Grafen's results are consistent with received evolutionary theory. However, they also suggest that the results are somewhat limited in scope. This takes us to the second question: do Grafen's results justify the practice of those adaptationist biologists who assume fitness-maximization in their day-to-day work?

The adaptationist credo that evolved phenotypes will appear designed to maximize an organism's fitness (however defined) is usually thought to be predicated on three assumptions: first, that natural selection is the major driver of evolutionary change; second, that the range of available phenotypes from which natural selection chooses is not heavily constrained; and third, that genetic architecture will not frustrate the fixation of the optimal phenotype in the population. These assumptions are the main foci of Gould and Lewontin's (1979) critique of adaptationism, for example.

Now Grafen's analysis does not address the relative importance of selection versus other evolutionary forces, such as drift. However, Grafen could reasonably reply that he is interested in the conditional proposition that *if* natural selection is the sole

[8] This point is made by Grafen (2006); however, Grafen (2014a) downplays the difference between maximization of individual and mean population fitness, suggesting that his project is continuous with earlier theorists' interest in mean fitness maximization. See Okasha and Paternotte (2014) and Birch (2016) for discussion.

driver of evolutionary change, then well-adapted organisms will evolve. Whether the antecedent of this conditional is true is an empirical issue; clearly one could not expect a formal argument to settle the matter.

What about the range of available variation? The point to note here is that Grafen's set of possible phenotypes, on which the objective function is defined, is a purely abstract set. Nothing is assumed about how big this set is, nor how different the phenotypes are from each other, nor how different their fitnesses. Thus Grafen's analysis is compatible with the presence of strong developmental constraints on the available phenotypic variation or with the complete absence of such constraints. But adaptationist reasoning is premised on the assumption that developmental constraints do not greatly limit the available variation; for if they did, then well-designed phenotypes could not evolve. Though this is an important point, I think that there is a plausible reply open to Grafen, namely that he is trying to vindicate one strand of adaptationist reasoning: the assumption that natural selection will lead the best available phenotype to evolve. This in itself would be a significant result, and again, it is perhaps unreasonable to expect a formal analysis to do more.

Turning to genetic architecture, consider the simple case of heterozygote superiority, or overdominance. Suppose there are two alleles *A* and *B* at a locus, and thus three genotypes *AA*, *AB*, and *BB*; genotype determines phenotype directly, in such a way that genotype viabilities satisfy $w_{AB} > w_{AA} = w_{BB}$; and mating is random. Since the *AB* heterozygote cannot breed true, selection will not lead the optimal phenotype to be fixed in the population: homozygotes will be produced anew each generation. Adherents of fitness-maximization are therefore making the empirical bet that such situations are fairly rare, or that natural selection will eventually manage to alter the genotype-phenotype map to circumvent the problem. But Grafen's selection-optimality links hold true in the case of heterozygote superiority, whether or not the genotype-phenotype map is malleable.

To see this, note that gene frequencies in this example will be stable across generations at $p = \frac{1}{2}$. However, this does not count as a selective equilibrium in Grafen's terms, since there is still 'potential for selection': a dominant mutant coding for the optimal phenotype would spread if it arose. So Grafen's links 1 and 4 both hold true, since their antecedents are both false. It is false that all individuals are optimal, so link 1 is true; and it is false that selective equilibrium obtains, so link 4 is true. Hence despite the fact that the optimal phenotype fails to fix in the population, all of Grafen's links hold true. This is all very well; but it is hard to see how it vindicates the assumption of fitness-maximization. The latter is usually taken to mean that genetic constraints can be safely ignored, that is, that the optimal phenotype will actually evolve; whereas Grafen's links hold true even in the case where genetic constraints prevent this from happening.

To be fair, Grafen is well aware of this point. He describes cases in which individual fitness maximization does not result but his links still hold true, such as the heterozygote superiority case, as 'exceptional', arguing that his results show that individuals having fitness-maximizing phenotypes is the 'general expectation' in biology (2014a, p. 12). But it is unclear on what basis this judgment is made. After all, non-additive interactions in general are quite common, between and within loci, so fitness non-additivity cannot be assumed to be negligible; moreover, gene frequencies need not reach equilibrium at all, as in principle they may cycle indefinitely, as noted previously. It might be argued that the success of adaptationist reasoning in biology shows that fitness non-additivity does not, in fact, prevent optimal phenotypes from evolving. This is plausible; but the adaptationist's assumptions would then be justified by their empirical success, not by Grafen's links.

Grafen (2014a) addresses this issue by introducing a contrast between one of his selection-optimality links holding 'trivially' and 'substantively', where the former means that the link holds because its antecedent is false. He suggests that in the problem cases where fitness is not maximized, such as heterozygote superiority, some of the links only hold true trivially; and that the assumptions of working biologists might be met if the links held true 'in a significant way' (p. 12). But note that in every case, two out of three of links 1, 2, and 3 must hold true trivially, since their antecedents constitute a partition of logical space (i.e. exactly one of the antecedents is true in every possible circumstance). So Grafen's links cannot all hold true substantively. It seems, therefore, that the trivial/substantive distinction does not help characterize the 'exceptional' cases that adaptationist biologists assume to be rare.

The upshot, I think, is that Grafen's selection-optimality links do not quite do the job that he asks of them. The links establish a connection between natural selection and optimality; and it is true that adaptationist biologists assume a connection between these two things. But the connection that the latter assume is not quite the one that Grafen's links establish. In effect, adaptationists assume that if natural selection is the only evolutionary force at work, it will lead to well-designed phenotypes; but this is stronger than what Grafen has shown. Grafen might reply that this stronger selection-optimality connection is not always true, so cannot possibly be proved mathematically. I think this is correct; but even so, given the distance between what Grafen's results prove and what adaptationists assume, it is hard to see why his project constitutes a vindication of the adaptationists' assumptions rather than a demonstration that they are only entitled to weaker ones.

Despite these criticisms, I see much of value in Grafen's project. I share his view that talk of agents, goals, and objectives is a natural way to capture the Darwinian notion of organismic design, and is implicit in much biological practice. I think Grafen is right to emphasize the conceptual distance between natural selection and phenotypic adaptation, and to ask how it can bridged. I think that Grafen's strategy of 'endogenizing' the fitness function, by defining it as whatever satisfies his selection-optimality links, constitutes an important move. Grafen's project offers a striking vindication of the unity-of-purpose constraint on agency, as we have seen. Moreover, I think that Grafen's selection-optimality links are very plausibly *necessary* conditions for individuals to be treated as akin to agents with goals. For the truth of links 1 and 4, in particular, is necessary if selection is to lead to fixation of the optimal phenotype in the population, which in turn is necessary if all individuals are to behave as if trying to maximize the objective function. The foregoing criticisms show only that Grafen's links are not sufficient conditions, since in some cases the links hold true even though adaptationist reasoning, and thus agential thinking (type 1), fails.

4.3 Frequency-dependent Selection

The analyses of Wright, Fisher, and Grafen focus on frequency-independent selection, the traditional home for ideas about optimization in evolution. However, contemporary biology recognizes that frequency-dependence is widespread, even perhaps ubiquitous.[9] Traits that affect resource competition, predation, and mating success, for example, typically give rise to situations where a type's relative fitness depends on its frequency in the population. In Maynard Smith's canonical example, an organism playing 'hawk' will flourish in a population full of passive 'doves'; but a dove will do better in a population composed mainly of hawks (Maynard Smith 1982). What if anything can be said about natural selection's 'optimizing tendency' in this context?

Frequency-dependent selection is a large topic, so here I focus only on core issues. Let us ask firstly what optimization or fitness-maximization means in this context. It is clear that selection need not increase mean population fitness $\overline{w}$ when the fitness of the competing types is frequency-dependent, as noted earlier. For even if an individual's traits are transmitted with perfect fidelity to its offspring, its fitness value will not be—since it depends on the trait-distribution in the population, which is evolving. Thus under frequency-dependence, we should obviously not expect natural selection to increase mean fitness over generations, a point that was recognized as early as Haldane (1932).

What about individual optimization? Here an ambiguity should be noted. Some authors understand the statement that evolution optimizes individuals' traits narrowly, to mean optimization of a one-placed function mapping individual trait value onto fitness (e.g. Diekmann 2004). In this sense, optimization only occurs when selection is frequency-independent (and even then not always). However, others take optimization to include 'best response', that is, having a trait that maximizes individual fitness conditional on the trait distribution in the rest of the population (e.g. Abrams 2001). In this sense, frequency-dependent selection may lead to optimization, for example, if it leads to an evolutionary stable strategy (ESS) in the population. For by definition, a trait is an ESS when it is uninvasible by mutants when fixed in a population, and this requires (at least) that the trait is a best response to itself. Thus at evolutionary stability, as defined by the ESS criterion, each individual is maximizing its fitness conditional on the state of the rest of the population.

This issue is largely semantic; however, I favour including best response as part of optimality, in order to preserve the conceptual link between adaptation and

[9] Dieckmann and Ferriere (2004) argue that frequency-dependence is likely to be the norm in evolution, but is given short shrift in textbooks.

optimality. Traits that evolve by frequency-dependent selection do admit of adaptationist explanation, after all. Take for example the fifty-fifty sex ratio found in many out-breeding species; as is well-known, this is the stable equilibrium of a selective process in which there is an advantage to any individual in producing more of the rarer sex. We can thus explain why a species' evolved sex-determination mechanism yields equal proportions of males and females in terms of the advantage that this mechanism confers on an individual conditional on the rest of the population having it too. This is a *bona fide* adaptationist explanation, yielding the unique type of understanding characteristic of such explanations.

In early work on frequency-dependent selection, for example, Maynard Smith (1982), where the focus was on finding equilibria rather than studying dynamics, it was often simply assumed that natural selection would lead a population to an ESS (possibly in mixed rather than pure strategies). If this assumption were true, it would mean that frequency-dependence is fully compatible with the Darwinian notion that natural selection leads to adaptation or optimality, providing that the latter is defined so as to include best-response; and indeed this was how many researchers saw it (Maynard Smith 1982, Mitchell and Valone 1990). However, more recent work increasingly casts doubt on this assumption. Dynamical models show clearly that when fitness is frequency-dependent, then even when natural selection is the only factor affecting a population, non-adaptive (i.e. non-ESS) outcomes are entirely possible (Abrams 2001).

This point comes across clearly in the field of adaptive dynamics.[10] This is an approach to frequency-dependent selection which aims for greater ecological realism than traditional evolutionary game theory. The focus is on continuous phenotypic traits, which are assumed to breed true—thus genetic constraints are ignored. Mutations of small effect arise periodically, on an evolutionary timescale. It is assumed that a population is at demographic equilibrium (neither growing nor shrinking) and is monomorphic whenever a new mutation arises. The fate of a mutant depends on whether it enjoys a positive or negative growth rate in the environment set by the resident. This is captured by the 'invasion fitness function' $w(x, y)$, defined as the growth-rate of a rare mutant with trait value y in a population monomorphic for trait value x. (Assume that the trait is one-dimensional, so both x and y are real numbers.) This growth-rate depends, of course, on how the mutants fare, in terms of survival and reproduction, compared to the residents.

Since the mutation is of small effect, that is, x and y are very close, the fate of the mutant depends on the sign of the fitness gradient in the vicinity of the resident trait x, which is defined as:

$$G(x) = \left.\frac{\delta w(x, y)}{\delta y}\right|_{y=x} \qquad (4.2)$$

[10] See for example Geritz *et al.* (1998), Metz (2012), Diekmann (2004), and Doebeli (2011).

This is highly intuitive. For if $G(x) > 0$, then mutants with $y > x$ do better than the resident, in the environment set by the resident, so can invade; while if $G(x) < 0$, then mutants with $y < x$ can invade. However, this does not tell us what will happen when $G(x) = 0$, that is, if the population is at a point in phenotype space where the fitness gradient vanishes. Such a point is known as a 'singular point' (or singular strategy) and represents a potential endpoint of the evolutionary process. Under the assumptions of adaptive dynamics, a series of trait substitutions occurs until a population reaches a singular point x^* at which $G(x^*) = 0$. Note that at such a point, the invasion fitness function $w(x^*, y)$ may be either a local maximum or minimum with respect to the mutant trait y.[11]

A central conceptual point of adaptive dynamics is to distinguish two senses in which a singular strategy x^* may be stable.[12] The first, known as evolutionary stability or uninvasibility, means that if x^* is fixed in the population, it cannot be invaded by nearby mutants; this is the traditional ESS concept, and simply requires that the singular point defined by $G(x^*) = 0$ should be a maximum, that is, any mutants do worse than the resident. The second, known as convergence stability, means that the population will evolve towards the singular point from nearby points in phenotype space that is, the trait x^* is an attractor of the evolutionary dynamics. Both of these stability concepts can be characterized in terms of the partial derivatives of the invasion fitness function $w(x, y)$ (Box 4.2).

Box 4.2. Uninvasibility and convergent stability

Uninvasibility requires that a singular point x^*, at which $G(x^*) = 0$, is a local maximum of $w(x^*, y)$:

$$\left.\frac{\delta^2 w(x,y)}{\delta y^2}\right|_{y=x=x^*} < 0 \qquad (4.3)$$

Convergence stability requires that the population evolves towards x^* from nearby points, that is, if $x < x^*$ then $G(x) > 0$, and if $x > x^*$ then $G(x) < 0$. This requires $G'(x^*) < 0$. By the chain rule, this gives:

$$\left.\frac{\delta^2 w(x,y)}{\delta y \delta x}\right|_{y=x=x^*} + \left.\frac{\delta^2 w(x,y)}{\delta y^2}\right|_{y=x=x^*} < 0 \qquad (4.4)$$

When inequality (4.4) is satisfied but (4.3) is not, x^* is convergence stable but not uninvasible, and $w(x^*, y)$ is at a local minimum.

[11] Here we ignore the degenerate case in which the second derivative of the invasion fitness function vanishes.

[12] This distinction is originally due to Eshel (1983).

The notions of convergence stability and uninvasibility are independent in both directions: neither implies the other. If a singular point x^* is uninvasible but not convergence stable, this means that were the population to reach that point it would be resistant to invasion by mutants; however, evolution will not lead it there from nearby points. Conversely, if a singular point is convergence stable but not uninvasible, this means that evolution will drive the population to that point from nearby points, only for it to become susceptible to invasion by mutants. As mutants invade, the resident trait value moves away from x^*, but this changes the selective pressure in such a way that the population is pushed back again towards x^*. At such a point x^* in phenotype space, the fitness function $w(x^*, y)$ is at a minimum with respect to y, and yet evolution by natural selection drives the population there! Such a point is known as a 'branching point' of the dynamics, since a population at the point can be invaded by simultaneous mutants on either side of the resident strategy x^*, leading it to experience disruptive selection and split into two phenotypically divergent sub-populations.

The intuitive explanation of how convergence to a fitness minimum can occur is that the individual fitness surface, described by the local fitness gradient, is continually changing. (Note that the local fitness gradient in equation (4.2) is always evaluated at the resident trait.) More precisely, at each point in phenotype space, that is, each possible resident trait value, evolution proceeds in accordance with the gradient at that point, favouring mutants with a positive growth rate. When a mutant invades and goes to fixation, this changes the resident trait value, *which in turn may cause the invasion fitness function to shift*, meaning that the fitness gradient itself may change. If the effect on the fitness gradient is large enough, measured by the first term on the left of inequality (4.4), then even if the second term is positive, that is, x^* is a minimum, the inequality may be satisfied. If so, then the singular point x^* is convergence stable and so the population is driven towards it.

This process is illustrated in Figure 4.1. The singular point x^* is a minimum of the invasion fitness function $w(x^*, y)$—so mutants with either higher or lower trait values can invade. (Note that by definition, x^* has zero growth rate in the environment in which it is itself resident, that is, $w(x^*, x^*) = 0$, since the population is at demographic equilibrium; thus the graph of $w(x^*, y)$ is tangent to the horizontal axis.) However, despite x^* being invasible, evolution will drive the population to x^* from nearby points, since the fitness gradient is positive for all $x < x^*$, and negative for all $x > x^*$. Thus for example, suppose that the resident trait value is x_2, so that the invasion fitness function is $w(x_2, y)$. The slope of $w(x_2, y)$ at the point $y = x_2$ is positive, as the dotted line shows, which means that mutants with trait value greater than x_2 will have a positive growth rate so the population evolves towards x^*. Conversely, if the resident trait value is greater than x^*, then mutants with a lower trait value are favoured, so again the population evolves towards x^*. In this way natural selection leads the population to converge to x^* from nearby points, despite x^* being a minimum, not a maximum, of the fitness function.

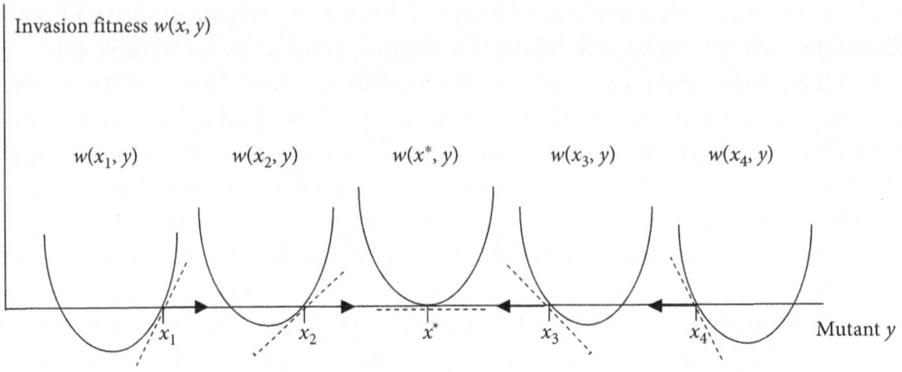

Figure 4.1. Convergence to a fitness minimum
Source: redrawn from Doebeli (2011), p. 17, by permission of Princeton University Press

The fact that natural selection can lead to fitness minimization is rather counter-intuitive, conflicting with our everyday understanding of how natural selection is supposed to work. For note that 'fitness', in the invasion fitness function, refers to individual fitness not mean population fitness. So the point is *not* the obvious one that with frequency-dependence, selection may drive the mean population fitness $\overline{w}$ down. Rather, the point is that selection may lead to fixation of a trait in a population which minimizes the fitness of individuals bearing the trait, conditional on the population state. Pre-theoretically one might have thought this impossible—since selection always favours fitter individuals, how can it lead to fitness minimization? However, this reaction, though understandable, ignores the feedback effect on the environment, manifest in the fitness gradient changing as the population evolves. That is, our naive understanding of Darwinian evolution, as a process with an inherent tendency to produce well-adapted individuals, does not hold water under some forms of frequency-dependence.

The empirical significance of branching points is a controversial topic, which I do not wish to enter into.[13] Rather I want to stress a conceptual point. On the face of it, adaptive dynamics makes assumptions which are favourable to the adaptationist viewpoint. Strategies are assumed to breed true, so genetic constraints are ignored; a continual supply of mutation is assumed; populations are assumed to have reached equilibrium; environmental fluctuations are ignored; and natural selection is the only evolutionary force at work. Thus one might have thought that well-adapted phenotypes, in the sense of ones that maximize individual fitness conditional on their being fixed in the population, would inevitably evolve. However, this is not the

[13] One issue here is whether branching points are a likely explanation of sympatric speciation. See Doebeli (2011).

case. The population may be driven to a point at which such conditional fitness-maximization obtains, that is, a convergence stable and uninvasible point in phenotype space, but not necessarily. That attractors of the dynamics exist at which individual fitness is minimized shows that frequency-dependent selection can yield outcomes that are highly non-adaptive, not just at the population level but also at the individual level. So despite its name, 'adaptive dynamics' does not necessarily produce what would ordinarily be thought of as adaptation.

The upshot, therefore, is that with frequency-dependence, we cannot assume that natural selection has an inherent optimizing tendency, even if we define 'optimal' to include best-response. That is, even if we replace the simplistic idea that natural selection adapts organisms to a static environment with a more nuanced view in which the environment co-evolves, we still cannot conclude that natural selection has any inherent tendency to produce adaptation.

There are two points of philosophical significance here. The first concerns what we have called agential thinking (type 2), the metaphorical assimilation of the *process* of natural selection to an agent's deliberate choice. We argued in section 1.4 that this metaphor can be of value but is potentially misleading, since it would only really apply well if Darwinian evolution had an inherent directional tendency, such as tending towards maximum fitness, and thus mimicked a goal-directed process. Adaptive dynamics teaches us that this is not so: in principle, selection can fail to maximize fitness, even in the absence of other forces. Again, this shows the inauspiciousness of agential thinking (type 2) as a heuristic for thinking about natural selection. It is as if 'mother nature' continually strives for a goal but, in some cases at least, ends up no nearer it. However, this leaves untouched agential thinking (type 1)—the practice of treating a well-adapted organism as akin to an agent with a goal, for this is a way of thinking about adaptation, not selection. This point is elaborated further in section 4.4.

The second moral concerns the prospects of applying an argument akin to Grafen's to the case of frequency-dependent selection. Recall Grafen's aim of establishing a mathematical link between the dynamics of natural selection and the optimality of individuals' phenotypes. Intuitively the existence of such a link, under frequency-dependence, would require that the evolutionary dynamics carry the population to a point at which individuals have phenotypes that maximize their fitness conditional on the rest of the population. We have seen that while this may happen, it need not: convergence to a fitness minimum is also perfectly possible. Adaptive dynamics thus teaches us that a selection-optimality link of this sort does not hold true in general, under frequency-dependence.

A weaker selection-optimality link does exist, however, which is perhaps a more direct analogue of the links that Grafen postulates. In particular, it is quite true that *if* a population is at a singular point in phenotype space such that all individuals have phenotypes that maximize the invasion fitness function conditional on the resident trait, then no mutants can invade, and vice-versa. But it does not follow that evolution will lead the population to such a point, nor that it will 'tend' to, even if natural selection is the only evolutionary force at work. As with the selection-optimality links that Grafen establishes in the frequency-independent case, this seems too slender a basis on which to claim that selection has any inherent tendency to produce adaptation.

4.4 Empirical or Theoretical Justification?

Let us stand back and reflect. In this and the previous chapter, we have examined Wright's, Fisher's, and Grafen's attempts to formalize and justify Darwin's argument that natural selection will lead to adaptation, or apparent design, in nature. We have argued that none of the three is entirely successful, though each contains important insights; and that frequency-dependence provides further reason to doubt that natural selection will automatically yield adaptation, even when the definition of adaptation is suitably broadened to include conditional fitness-maximization. If this is correct, it suggests a general moral: that adaptationism in biology must ultimately be justified on empirical rather than theoretical grounds.

This may seem obvious; however, it conflicts with the widespread assumption that natural selection has an intrinsic tendency to produce adaptation, unless constraints or non-selective forces intrude. (The very language of 'constraints' seems to imply such a tendency.) I think this assumption has two sources. The first is simply the logic of natural selection itself, which in its purest form involves the proliferation of better-adapted types at the expense of poorer ones. Surely, therefore, cumulative rounds of natural selection must produce organisms that are well-adapted to their environment, unless other factors intrude? However, this argument should not tempt us. The complexities of genetics, inheritance, frequency-dependence, population structure, and stochasticity all mean that in reality, evolution by natural selection need not assume this simple form, and often will not do.

The second source is an assumption about scientific explanation. The existence of adaptation is an empirical fact; following Darwin, contemporary biologists believe that natural selection on heritable variation is the explanation. If this is a good explanation, as is widely believed, it is tempting to assume that natural selection must *guarantee* that adaptation will result, at least given certain assumptions, or at a minimum render it highly likely. For many scientific theories explain a phenomenon precisely by showing that the phenomenon had to occur, or was very likely to occur, given the truth of the theory; indeed Hempel (1965) regarded this as the hallmark of a good scientific explanation. Thus we are led to think that there must be a theoretical principle that says that natural selection will result in well-adapted organisms, at least *ceteris paribus*. Maynard Smith (1978) gives voice to this conviction when he says that adaptation is both observed fact and 'a necessary consequence of natural selection' (p. 38).

I suspect that Grafen (2014b) is motivated by a similar consideration when he argues, in response to critics, that if his selection-optimality links are too weak to capture Darwin's argument in *The Origin*, then we should conclude that 'Darwin was wrong', since those links represent 'the best we can do to defend Darwin's idea that the mechanical processes of inheritance and reproduction can give rise to the appearance of design' (p. 290). However, I do not think we need to go this far. Darwin's argument describes a conceivable mechanism by which well-adapted organisms could have arisen—cumulative natural selection—and he makes the empirical claim that this is how adaptation did in fact come about. I suggest that we can accept this without embracing the idea that natural selection must lead to this outcome, or even makes it probable, and thus without being committed to the existence of a general theoretical principle to the effect that selection leads to fitness-maximization, or adaptation.

This requires us to reject the idea that a good scientific explanation must render a phenomenon probable in an absolute sense; but that principle is arguably too strong anyway.[14] A more plausible principle is that a good explanation must render the phenomenon it explains more probable that it was before. The difference here is between the conditions:

$$\Pr(e|T) > c \tag{1}$$

$$\Pr(e|T) > \Pr(e) \tag{2}$$

where T is the theory doing the explaining, e is the phenomenon to be explained, and c is some fixed threshold, for example, 0.9. Note that conditions (1) and (2) are logically independent. I suggest that (2) is necessary for a theory to explain the phenomenon but (1) is not—though it is clearly a desirable feature. In the case at hand, T is the theory of evolution by natural selection and e is the existence of adaptation, or apparent design. Our conclusion above—that there is no theoretical principle to the effect that natural selection must or will probably produce adaptation—implies that condition (1) is not satisfied, or at least has not been shown to be. However, condition (2), which is equivalent to $\Pr(e|T) > \Pr(e|\neg T)$, is satisfied. For $\Pr(e|\neg T)$ is surely very low—as the existence of adaptation would be miraculous if the Darwinian explanation were not correct, given that no other plausible naturalistic explanation has ever been offered. I suggest, therefore, that adaptation is a possible rather than a necessary consequence of natural selection, but that this does not impugn the validity of Darwin's argument.

Finally, let us consider the implications for the use of agential concepts in evolutionary biology. Recall our previous distinction between the two types of agential thinking. Type 1 involves treating an evolved organism as akin to an agent with a goal towards which its phenotypic traits conduce; type 2 involves analogizing the process of natural selection to a process of deliberate choice between alternatives. We argued that these two modes of agential thinking are related though distinct: the

[14] See Salmon (1989) for numerous counterexamples.

former is a way of thinking about adaptation (the product), the latter about natural selection (the process). Our conclusion above reinforces the importance of the type 1/type 2 distinction. The two biological phenomena—selection and adaptation—to which agential concepts have been applied are related contingently, not as a matter of logical necessity.

We have found reason to regard agential thinking (type 2) with suspicion. Given that natural selection does not always lead towards higher fitness, it does not mimic a rational, goal-directed process; so the anthropomorphic assimilation of natural selection to agential choice is not in general advisable. However, this does not undermine the validity of agential thinking (type 1) as a way of understanding adaptation. Rather, it suggests that the justification for this mode of thinking is empirical rather than theoretical. It is not the logic of natural selection itself, for example, that it involves fitter types being 'chosen' by mother nature at the expense of the less fit, that justifies treating an evolved organism as akin to an agent with a goal. Rather, it is the empirical fact that contemporary organisms are often well-adapted, with traits whose evolved functions are (for the most part) complementary rather than antagonistic, and which can be explained as contributions to the organism's goals of survival and reproduction, that allows them to be treated usefully as agent-like.

4.5 Conclusion

This chapter has delved further into the issue of fitness-maximization in biology, building on the previous chapter's examination of the classical treatments by Wright and Fisher. We have focused on Grafen's maximizing agent concept, and his aim of linking the process of natural selection with the optimality of individuals' phenotypes. We have found that Grafen's analysis is strikingly consonant with our own unity-of-purpose constraint on agency; however, it falls short of vindicating the Darwinian idea that natural selection will lead individuals to have fitness-maximizing phenotypes. The truth of Grafen's selection-optimality links is a necessary though not sufficient condition for that conclusion to be justified. We then turned to frequency-dependent selection, as studied in the modern adaptive dynamics literature, and found a similar moral. Natural selection does not necessarily lead to optimal phenotypes, in the sense of ones which maximize individual fitness conditional on the rest of the population, to evolve. The overall lesson is that there is no theoretical principle to the effect that natural selection will tend to produce adaptation, contrary to what is often thought. The justification for adaptationism, and thus for agential thinking (type 1), must ultimately be empirical.

5

Social Evolution, Hamilton's Rule, and Inclusive Fitness

5.1 Introduction

Social evolution is the study of how natural selection shapes an organism's social behaviour, understood broadly to mean any organismic trait that affects the fitness of others. Of particular interest in this field are altruistic traits, which impose a fitness cost on the bearer but benefit others. Darwin himself realized that such traits are a puzzle, since at first blush it is hard to see how they could evolve; he sketched a possible resolution of the puzzle that anticipated later developments.[1] Our focus here will be on inclusive fitness theory, the widely used framework for understanding social evolution devised by W. D. Hamilton (1964) and the subject of much recent controversy.

Inclusive fitness is a refinement of the traditional Darwinian fitness concept—number of surviving offspring—designed for use in a social context. Hamilton realized that where social interactions occur between relatives, the effect of a trait on its bearer's Darwinian (or personal) fitness is not the right test of whether it will spread. A gene encoding an altruistic behaviour, which causes a reduction in the number of offspring of its host organism, may be favoured by natural selection if the behaviour sufficiently enhances the reproductive success of the organism's relatives—since relatives share genes, hence the notion of 'kin selection'. Essentially, Hamilton's idea was to define a new fitness measure which takes account of such indirect effects and thus supplies a correct criterion for when a social trait will be favoured by natural selection.

The main qualitative prediction of inclusive fitness theory is that altruism should be more common among relatives than non-relatives. Empirically this prediction has been confirmed in diverse taxa including insects, vertebrates, and microbes, where genetic relatedness has provided the key to understanding observed patterns of behaviour. However, inclusive fitness is not the only way to study social evolution, and has its share of critics as well as supporters.[2] Points of controversy include

[1] Darwin's discussion of 'selection applied to the family' in the *Origin* anticipates kin selection (1859, p. 258), while that of the evolution of 'moral faculties' in *The Descent of Man* anticipates multi-level selection (1871, p. 166).

[2] See in particular Nowak *et al.* (2010), Abbot *et al.* (2011), and Allen and Nowak (2015). This controversy is analysed by Birch (2014) and Birch and Okasha (2015).

how generally applicable inclusive fitness theory is, whether it relies on problematic assumptions, and whether it is methodologically superior to other approaches. Interestingly, this controversy relates directly to the use of agential thinking (type 1) in evolutionary biology. For as we shall see, much of the attraction of the inclusive fitness concept is that it suggests a 'goal' for social behaviour, promising to bring such behaviour within the ambit of the organism-as-rational-agent heuristic described in previous chapters. However, if the critics are right, this promise is illusory.

This chapter has two aims. The first is to offer a philosophical analysis of the ongoing controversy over inclusive fitness theory. The second is to relate the inclusive fitness concept to our previous treatment of agential thinking in chapter 1, and fitness-maximization in chapters 3 and 4. The structure of this chapter is as follows. Section 5.2 introduces Hamilton's rule and the concept of inclusive fitness maximization, emphasizing the distinction between them. Section 5.3 sketches a simple evolutionary model of social behaviour with additive payoffs, and shows that at equilibrium, organisms will behave like rational agents trying to maximize their inclusive fitness. Section 5.4 considers the same model with non-additive payoffs, describes its dynamics using a generalized version of Hamilton's rule, and asks what becomes of the organism-as-rational-agent notion. Section 5.5 considers the causal meaning of the generalized Hamilton's rule, and explores a link with Fisher's fundamental theorem. Section 5.6 concludes.

5.2 Hamilton's Rule and Inclusive Fitness Maximization

Hamilton's original formulation of inclusive fitness theory contained two distinct though related aspects, not sharply distinguished from each other. The first is Hamilton's rule $rB > C$, the back-of-the-envelope criterion for when a gene that causes a social behaviour will be favoured by natural selection. Here C is the fitness cost to the actor, B the benefit to the recipient, and r the 'coefficient of relationship' between them, which measures how closely related they are. This aspect of the theory fits with the gene's eye view of evolution, encouraging us to treat altruism as a gene's way of helping copies of itself in its host's relatives. The second is maximization of inclusive fitness, rather than personal fitness, as the 'goal' towards which an individual's social behaviour will appear designed.[3] This aspect fits with the traditional individualist view of evolution, and is frequently employed by behavioural ecologists seeking to understand observed behaviour in the field.

Where costs and benefits make additive contributions to fitness, as assumed by Hamilton (1964), these two aspects will typically dovetail. That is, the criterion for whether a gene for the social trait spreads is $rB > C$, and the effect of the trait on an

[3] This maximization claim should be distinguished from the claim that natural selection will maximize the average inclusive fitness of the whole population; the latter claim is found in Hamilton (1964), the former in Hamilton (1970).

individual's inclusive fitness is $(rB - C)$, a quantity known as the 'inclusive fitness effect'. Thus the social trait spreads if and only if the inclusive fitness effect is positive. As long as $(rB - C)$ stays constant over generations, this suggests that natural selection will drive the population to an equilibrium in which individuals have traits that maximize their inclusive fitness. But with non-additivity the two aspects of Hamilton's theory can come apart, as we shall see.

The relation between the two aspects is not fully settled. Much theoretical work has focused solely on the first aspect, and indeed the second is often omitted from expositions of kin selection theory altogether. However, a number of recent authors have argued for the central importance of inclusive fitness maximization as the 'goal' of individual behaviour, on the grounds that it allows social behaviour to be brought within the traditional adaptationist paradigm (Grafen 2006, Queller 2011, Gardner *et al.* 2011). This is the main conceptual advantage of inclusive fitness theory, they argue, and provides a strong reason to favour it over alternative approaches to social evolution, such as multi-level selection.

In terms of our previous discussion, these authors are suggesting that the inclusive fitness concept is valuable because it allows us to apply agential thinking (type 1) to social behaviour. That is, we can treat an organism as akin to a rational agent trying to achieve a goal though its choice of social action; so inclusive fitness plays the role of the agent's utility function. This is certainly an attractive idea; however, its status is unclear for two (related) reasons. First, the most careful attempt to defend the idea, due to Grafen (2006), is explicitly restricted to the case where costs and benefits are additive. This is a significant limitation given that synergistic interactions are common in nature.[4] So our first task is to ask how widely the organism-as-rational-agent concept, with utility equated to inclusive fitness, can be applied in a social evolutionary context. Second, critics have argued that the whole edifice of inclusive fitness theory rests on assumptions which greatly limit its applicability. Both aspects of the theory have been criticized. Thus Nowak *et al.* (2010) argue that Hamilton's rule 'almost never holds' (p. 1059); while Allen and Nowak (2015) argue that social evolution will not generally lead to maximization of inclusive fitness (p. 20138). However, both of these claims have been vigorously disputed.[5] Our second task is to examine this controversy.

5.2.1 *Preliminaries*

What exactly is inclusive fitness? Hamilton's original definition was:

> the personal fitness which an individual actually expresses ... once it is stripped of all components which can be considered as due to the individual's social environment ... then augmented by certain fractions of the quantities of harm and benefit which the individual himself causes to the fitnesses of his neighbours ... The fractions in question are simply the coefficients of relationship. (1964, p. 8)

[4] That costs and benefits are often non-additive is stressed by Frank (1998) and Lehmann and Rousset (2014).

[5] See in particular Abbot *et al.* (2010) and Gardner *et al.* (2011).

This definition sounds complicated but the underlying idea is simple, namely to re-assign all fitness components to the actors that cause them. A focal individual's personal fitness, that is, reproductive output, depends partly on its own phenotype and partly on the social actions of others. To compute their inclusive fitness, Hamilton tells us, we subtract the second component, and then add the effect of the focal individual's social actions on the personal fitness of others, weighted by how closely related they are. Therefore, the inclusive fitness of individual i is:

$$IF_i = w_i + \sum_j x_{ij} r_{ij} \tag{5.1}$$

where w_i is i's non-social fitness, x_{ij} is the effect of individual i's actions on the fitness of individual j, and r_{ij} is the relatedness of individual i to j. The summation is over all population members affected by individual i's actions.

Three related points about this definition deserve comment. First, the definition is framed in explicitly causal terms: the costs and benefits are the causal effects of a social action on fitness. Second, the definition assumes that social actions make additive contributions to fitness. For if not, it will not be possible to express an individual's personal fitness as the sum of asocial and social components, nor therefore to perform the 'stripping and augmenting' operation that Hamilton describes. To see this, suppose that a social action has a synergistic effect: the amount by which it increases a recipient's fitness depends on whether the recipient themselves performs the action. We cannot then separate the component of an individual's fitness due to social environment from the component due to its own phenotype: its fitness is a non-additive function of both. Third, inclusive fitness defined this way is a quantity that is under the 'control' of the actor, that is, the amount that an individual gets depends on their own actions, not those of others (Grafen 2006). Thus in principle, each individual could optimize its inclusive fitness independently of others. However, if actor's control fails, we need to consider best-response rather than optimization *senso stricto*.

The original definition of inclusive fitness is not always adhered to. Many authors, including Hamilton himself, have sometimes used a simpler definition, namely an individual's personal fitness plus r times the fitness of each of its social partners (neighbours). This definition thus omits the stripping of the component of an individual's personal fitness that is due to social environment, and it augments the individual's fitness by r times the entire fitness of its social partner, rather than the component caused by the individual's actions. Grafen (1982) argued that this leads to an unwelcome double-counting. However, one point in its favour is that it can be applied whether or not social actions have additive effects on fitness. I refer to this definition of inclusive fitness as 'simplified', by contrast with the 'original' definition of Hamilton (1964).

Finally, what of the coefficient of relationship r? In Hamilton's original work r was defined in genealogical terms, as the probability that an actor and its social partner share an allele that is 'identical by descent'. This yields the familiar values of $\frac{1}{2}$ for full sibs, $\frac{1}{2}$ for offspring, $\frac{1}{4}$ for grandoffspring, and so on. This allowed Hamilton to capture the idea, first suggested by Haldane (1955), that an individual should value its relatives' reproduction in proportion to how closely related they are. But as has long been recognized, what really matters is the genetic similarity between actor and recipient, whether due to genealogical relatedness or not (as Hamilton (1975) noted). Thus modern treatments define r in statistical terms, as a measure of the genetic correlation between social partners, which is the approach followed here.

5.3 The Case of Additive Payoffs

In this section and the next we study a simple model for the evolution of social behaviour, analyse it in inclusive fitness terms, then examine how the organism-as-rational-agent idea applies to it. The model is too simple to be of intrinsic interest, but serves to fix ideas and represents a best-case scenario for the view that social evolution leads to inclusive fitness maximization. The model is a variant on one that has been widely used in discussions on the logical status of inclusive fitness theory.[6] However, these discussions have focused exclusively on Hamilton's rule itself, rather than on the idea that individuals will evolve to maximize their inclusive fitness. Our contribution here is to analyse both of these aspects of the theory at once.[7]

Consider an infinite population of haploid asexual organisms, with non-overlapping generations. The organisms engage in pairwise social interactions in every generation. Organisms are of two types: altruists (A) and selfish (S). A types perform an action that is costly for themselves but benefits their partner; S types do not perform the action. Type is hard-wired genetically and perfectly inherited.

An organism's payoff from the social interaction depends on its own type and its partner's type, which may be correlated. Payoffs are interpreted as increases in lifetime reproductive fitness over a unit baseline. The social action is assumed to affect only the actor and their partner, thus local interaction and kin competition are assumed absent. An A type incurs a cost of $-C$ as a result of its action and confers a benefit of B on its partner, where $C > 0$ and $B > 0$; thus the game is a Prisoner's Dilemma.

We let $V(i, j)$ denote the payoff to an actor from playing i when her opponent plays j, where $i, j \in \{A, S\}$. Payoffs to the actor, referred to as 'personal payoffs', are shown in Table 5.1. (Since the game is symmetric, partner payoffs are not shown explicitly.) Note that payoffs are additive: an altruist alters their own payoff by $-C$ and their partner's payoff by B, irrespective of the type of their partner.

There are three pair-types in the population, AA, AS, and SS, whose relative frequencies in the initial generation are P, $2Q$, and R, respectively, where $P + 2Q + R = 1$.

6 For example, Queller (1984), Gardner *et al.* (2011), van Veelen (2009), Rousset (2015).
7 The material in this and the next section is based on Okasha and Martens (2016a).

Table 5.1. Additive case, personal payoffs

		Partner	
		A	S
Actor	A	$B - C$	$-C$
	S	B	0

Table 5.2. Pair-type frequencies

Pair-type	Frequency
AA	$P = p^2 + rp(1-p)$
AS	$2Q = 2p(1-p)(1-r)$
SS	$R = (1-p)^2 + rp(1-p)$

The overall frequency of the A type in the initial generation is $p = P + Q$. The change in p over one generation is denoted Δp.

The sign and magnitude of Δp depend on the rules by which the pairs are formed. If pairing is random, then the S type must be fitter overall, so Δp will be negative. However, if pairing is assortative then the A type may be fitter overall, for the benefits of altruistic actions then fall disproportionately on other altruists. A simple regression analysis yields a measure of the statistical correlation between social partners. We use the variable p_i to indicate an organism's own type and p'_i to indicate its partner's type; thus $p_i = 1$ if the ith organism is an A, $p_i = 0$ otherwise; and $p'_i = 1$ if the ith organism is paired with an A, $p'_i = 0$ otherwise. We then compute the linear regression of p'_i on p_i, given by $b_{p'p} = \text{Cov}(p', p)/\text{Var}(p)$, which is a standard way of defining the r term of Hamilton's rule. Henceforth we refer to $b_{p'p}$ as r.

An explicit expression for r can be written in terms of the pair-type frequencies P, Q, and R:

$$r = \frac{P - (P+Q)^2}{(P+Q)(R+Q)} \tag{5.2}$$

Conversely, we can express P, Q, and R in terms of r and p, as shown in Table 5.2. This helps clarify the meaning of r. Essentially, each organism has an r chance of being partnered with a clone and a $1 - r$ chance of being partnered with a randomly chosen population member.

We can also write the conditional probability than an organism (or 'actor') has a partner of either type, given its own type, in terms of r and p, as shown in Table 5.3. This allows r to be expressed as a difference in conditional probabilities:

$$r = \Pr(\text{partner is } A \mid \text{actor is } A) - \Pr(\text{partner is } A \mid \text{actor is } S)$$

Table 5.3. Conditional probabilities

Pr(partner is A | actor is S)	=	$p(1-r)$
Pr(partner is S | actor is S)	=	$1-p(1-r)$
Pr(partner is A | actor is A)	=	$r+p(1-r)$
Pr(partner is S | actor is A)	=	$(1-p)(1-r)$

It follows that r ranges from -1 (perfect disassortment) to $+1$ (perfect assortment); when pairing is random, $r = 0$.

5.3.1 *Evolutionary analysis*

In Appendix 5.1, we show that the change in p over one generation is given by:

$$\Delta p = (rB - C) \cdot \mathrm{Var}(p)/\overline{w} \tag{5.3}$$

where $\overline{w}$ is average population fitness and Var(p) is the variance of the indicator variable p_i. Since Var(p) is non-negative, this tells us that as long as $0 < p < 1$, the A type will increase in frequency in the population whenever $rB > C$, which is of course Hamilton's rule. Since B and C are fixed parameters of the payoff matrix, this condition for the spread of altruism is frequency-independent as long as r itself does not change as the population evolves. Constancy of r across generations will often be a reasonable assumption, for the pattern of assortment in the population, which is what r measures, may be determined by biological factors, for example, dispersal, which are independent of the social trait that is evolving.

With the constant r assumption, the outcome of the evolutionary process is easily determined. If $rB > C$ the A type will spread to fixation; if $rB < C$ the S type will spread to fixation; if $rB = C$ there will be no evolutionary change.

5.3.2 *Rational agent analysis*

To apply the organism-as-rational-agent concept, we transpose our evolutionary model to a rational choice context, as follows. We consider two rational players playing a symmetric game. Each player has two pure strategies, A and S. If a player plays a *mixed strategy* this means that they randomize over their pure strategies; thus π_A denotes the mixed strategy in which A is played with probability π_A and S with probability $1 - \pi_A$. The payoff to a mixed strategy is then simply its expected payoff.

Each player has a utility function which measures how desirable they find the possible outcomes of the game; we assume that both players have the same utility function. Each player's goal is to maximize their utility function. One possibility is that the utility function is given by the personal payoffs in Table 5.1, in which case we write $U(i,j) = V(i,j)$, where $U(i,j)$ is the utility a player gets from playing i when their partner plays j, where $i, j \in \{A, S\}$. There are other possibilities too, as we shall see.

Once the players' utility function has been specified, the next step is to seek the Nash equilibrium (or equilibria) of the game. (A Nash equilibrium is a pair of strategies, possibly mixed, one for each player, each of which is a best response to the other.) Standard game theory predicts that if the players are rational, they will end up at a Nash equilibrium of the game, at which each player maximizes their utility conditional on the other's action. We can then ask whether the Nash equilibria of the rational game correspond to the outcomes of the evolutionary process described above. If so, we can conclude that evolution will lead organisms to behave as if trying to maximize the utility function in question.

This is a natural way of applying the organism-as-rational-agent idea in the context of social interactions, and thus of using agential thinking (type 1) to understand evolved social behaviour. It differs from Grafen's (2006) formalization of the same idea, which posits 'links' between gene-frequency change and individual optimization, as discussed in section 4.2. Our approach allows recovery of Grafen's main result, and by taking optimization to include best-response, that is, optimal choice conditional on the other player's choice, extends easily to the non-additive case.[8]

5.3.3 *Utility as inclusive fitness*

One possibility is that an agent's utility function depends on their partner's payoff as well as their own. For example, suppose that an agent's utility for any outcome is given by the quantity: personal payoff plus r times partner's payoff, that is, $U(i,j) = V(i,j) + rV(j,i)$. This transformation was first suggested by Hamilton (1971), and has been discussed by many subsequent authors.[9] It is a natural formalization of the idea that an actor, in their social behaviour, will care about their partner's payoff, discounted by relatedness, as well as their own payoff. Applying this transformation to the personal payoffs yields Table 5.4.

Note that the payoffs in Table 5.4 correspond to the 'simplified' rather than 'original' definition of inclusive fitness. For the actor's payoff has not been stripped of the component that is due to the partner's altruistic action (B), and has been augmented by r times the partner's entire payoff, rather than the portion of that payoff that is

Table 5.4. Additive case, simplified IF payoffs

		Partner	
		A	S
Actor	A	$(B-C)(r+1)$	$-C+rB$
	S	$B-rC$	0

[8] A similar approach to the one used here is found in Alger and Weibull (2012) and Lehmann *et al.* (2015).

[9] These include Grafen (1979), Bergstrom (1995), Day and Taylor (1998), Taylor and Nowak (2007), and Martens (2017).

Table 5.5. Additive case, original IF payoffs

		Partner	
		A	S
Actor	A	$-C + rB$	$-C + rB$
	S	0	0

caused by the actor (also B). Applying Hamilton's original definition exactly would lead to the payoff matrix in Table 5.5.

Note that Table 5.5 derives from Table 5.4 by subtraction of the quantity $B - rC$ from the left column. In game-theoretic language, Table 5.5 is thus a 'local shift' of Table 5.4, which means that their Nash equilibria are necessarily identical (Weibull 1995). Therefore, if the players' utility function is given by Table 5.5, game theory predicts exactly the same outcome(s) as if it were given by Table 5.4. So although taking Table 5.4 as the definition of inclusive fitness involves an element of double-counting, in this context at least it is harmless.

Since r was defined in the evolutionary model as a measure of assortment, it may seem odd, conceptually, to allow the players' payoffs in the rational model to depend on it. However, this makes sense as long as r is a fixed parameter, as we are assuming, to which the population is adapted. Since an organism does not know the type of its partner, evolution can induce it to behave optimally by equipping it with a utility function which depends suitably on partner payoff. Thus although r was originally introduced as a measure of assortment, in the context of the rational agent model it can serve as a measure of how much one player values their partner's payoff. This duality in the interpretation of r has been emphasized by Frank (1998).

5.3.4 *Results*

Suppose first that the utility function is personal payoff (Table 5.1). It is easy to see that (S, S) is the only Nash equilibrium of the game, since S strongly dominates A, that is, each player does strictly better by playing S irrespective of their partner's choice. This familiar result shows that agential thinking (type 1) fails for this choice of utility function, since it would have us conclude that altruism can never evolve, which we know to be false.

What if the utility function is inclusive fitness, either simplified (Table 5.4) or original (Table 5.5)? In that case, we can show the following. If $rB > C$ then (A, A) is the unique Nash equilibrium; if $rB < C$ then (S, S) is the unique Nash equilibrium; if $rB = C$ then (A, A) and (S, S) are both Nash equilibria, as is every pair of mixed strategies, in which case game theory makes no prediction about the players' choices (see Appendix 5.1 for proof).

It follows that with additive payoffs, defining utility as inclusive fitness rescues the organism-as-rational-agent concept. The condition for the A type to evolve, $rB > C$, is identical to the condition for (A, A) to be the unique Nash equilibrium of the rational

Table 5.6. Evolution–rationality link with utility = inclusive fitness

$rB > C$	$\iff$	A evolves	$\iff$	(A, A) is unique Nash equilibrium
$rB < C$	$\iff$	S evolves	$\iff$	(S, S) is unique Nash equilibrium
$rB = C$	$\iff$	no evolution	$\iff$	all pairs of strategies, pure and mixed, are Nash equilibria

Table 5.7. Additive case, Grafen 1979 payoffs

		Partner	
		A	S
Actor	A	$(B - C)$	$-C + rB$
	S	$(1 - r)B$	0

game; and similarly for S (Table 5.6). This supports the idea that evolution will lead organisms to appear as if trying to maximize their inclusive fitness, just as Hamilton originally argued.

An equivalent perspective on the situation is this. The quantity $(rB - C)$ is the difference in a player's inclusive fitness payoff between playing A and S, irrespective of what its partner does, as is clear from Tables 5.4 and 5.5. Thus we can determine whether the A type will evolve by asking whether a rational agent, who wants to maximize their inclusive fitness, would choose A over S. (Thus we can use Sober's (1998) 'heuristic of personification' to determine the outcome of evolution.) In short, equating utility with inclusive fitness ensures that the rational agent's choice coincides with that of an organism whose behaviour has been optimized by natural selection; so agential thinking works fine.

5.3.5 A caveat: uniqueness

One important caveat is needed. In the rational choice model of the previous section, inclusive fitness is not the only choice of utility function that yields the $rB > C$ condition for action A to be chosen over S. In game theory, the utility function is only ever unique up to choice of origin and unit; so any positive linear transformation will leave all Nash equilibria of the game unchanged.[10] Furthermore, a local shift of the utility function, which involves adding a constant to any column of the payoff matrix, will also leave unchanged the Nash equilibria, as noted previously.

One local shift of the simplified IF payoff matrix (Table 5.4) is of particular interest. If we add the quantity $(rC - rB)$ to the left-hand column of Table 5.4, we get the matrix in Table 5.7.

The payoffs in Table 5.7 are related to the personal payoffs (Table 5.1) by the transformation $U(i, j) = rV(i, i) + (1 - r)V(i, j)$. This transformation was first suggested by

[10] That is, a transformation of the form $U' = aU + b$, where $a, b \in \mathbb{R}$, $a > 0$.

Grafen (1979) in a related context, hence the label. By contrast with the simplified IF payoffs, which involve adding r times partner's payoff to the actor's personal payoff, the Grafen 1979 payoffs involve taking an $(r, 1 - r)$ weighted average of the personal payoff that would accrue to the actor if their partner had chosen the same as the actor and if their partner had made the choice that they actually did. This implies that the value an agent places on an action depends not just on the actual personal payoff that the action brings, but also on the personal payoff that would have ensued had their opponent reciprocated and chosen the same action themselves.[11]

Since the Grafen 1979 payoff matrix (Table 5.7) is a local shift of the simplified IF payoff matrix (Table 5.4), the Nash equilibria of the resulting games are identical; thus the organism-as-rational-agent heuristic works equally well with either (since in both cases, $rB - C$ is the payoff difference between playing A and S). Therefore while our simple model has vindicated Hamilton's claim that evolution will lead organisms to behave as if trying to maximize their inclusive fitness, it is important to see that inclusive fitness (simplified or original) is not the unique quantity of which this maximization claim is true. Proponents of inclusive fitness maximization appear not to have considered this non-uniqueness issue.

5.4 Non-additive Payoffs

Do the above results generalize to the non-additive case? To assess this, we consider a modification of our evolutionary model in which the payoff to an A type paired with another A type is $(B - C + D)$ rather than $(B - C)$. So the parameter D quantifies the deviation from payoff additivity, or synergistic effect, when two A types are paired together; D can be either positive or negative (Table 5.8). The resulting payoff structure is sometimes referred to as a 'synergy game', and has been widely discussed in the literature.[12] Again, we assume that pairs of organisms are drawn from an infinite population to play the game; type is genetically hard-wired, mutation is absent, and reproductive success is proportional to payoff. As before, Δp denotes the change in frequency of the A type over a generation.

Table 5.8. Non-additive case, personal payoffs

		Partner	
		A	S
Actor	A	$B - C + D$	$-C$
	S	B	0

[11] Bergstrom (1995) refers to the Grafen 1979 payoffs as 'semi-Kantian payoffs', as they partially implement Kant's 'golden rule' of doing unto others as you would have them do unto you.
[12] See for example Queller (1984), van Veelen (2009), Gardner *et al.* (2011), Rousset (2015).

Unsurprisingly, $rB > C$ is no longer the condition for Δp to be positive, so it seems that Hamilton's rule no longer holds (though if the social action has weak effects on fitness, it may hold approximately). However, a number of authors have argued that with non-additive payoffs, it is possible to recover an exact version of Hamilton's rule by suitably re-defining the cost and benefit terms, as partial regression coefficients.[13] This point has recently been argued in detail by Gardner *et al.* (2011) in relation to the two-player synergy game; the next section follows their treatment.

5.4.1 Generalized Hamilton's rule

To derive the generalized version of Hamilton's rule, we write an individual's personal fitness w_i as a multiple regression on its own type (p_i) and its partner type (p_i'):

$$w_i = \alpha + b_{wp.p'}p_i + b_{wp'.p}p_i' + e_i \tag{5.4}$$

where α is baseline fitness, before the social interaction, $b_{wp.p'}$ is the partial regression of an individual's fitness on their own type, controlling for their partner's type, $b_{wp'.p}$ is the partial regression of an individual's fitness on their partner's type, controlling for their own type, and e_i is the residual (Figure 5.1). These partial regression coefficients are measures of the average effect, *sensu* Fisher (1930), of actor type and partner type, respectively, on the actor's personal fitness.

Following Hamilton (1964), instead of considering the effect on the actor's fitness of their partner's action $b_{wp'.p}$, we can consider the effect on their partner's fitness of the actor's action, denoted $b_{w'p.p'}$. (This is the well-known switch from 'neighbour-modulated' to inclusive fitness, in which all fitness effects are assigned to the actor.) Given the symmetry of the situation, these two partial regression coefficients are numerically identical (Taylor *et al.* 2007). We then denote the $b_{wp.p'}$ and $b_{w'p.p'}$ coefficients as $-c$ and b, respectively.

Importantly, the regression equation (5.4) can be fitted whether or not the true relation between w, p_i, and p_i' is linear. In the non-additive case the relation is non-linear (since $D > 0$), which implies that the partial regression coefficients $-c$ and

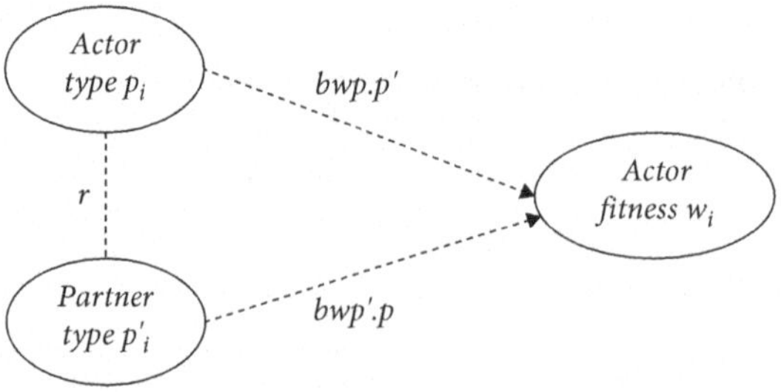

Figure 5.1. Direct and indirect determinants of fitness

[13] For example, Queller (1992, 2011) and Frank (1998).

b will be functions of population-wide gene frequencies, and liable to change as the population evolves. Therefore unlike C and B, which are fixed payoff parameters, c and b are population variables.

Following Gardner *et al.* (2007), we can write explicit expressions for c and b in terms of r, p, and the parameters of the payoff matrix B, C, and D. This yields:

$$-c = (-C) + (D) \cdot [r + p(1 - r)]/[1 + r] \tag{5.5}$$

$$b = B + (D) \cdot [r + p(1 - r)]/[1 + r] \tag{5.6}$$

5.4.2 *Evolutionary analysis*

Using the above definitions of c and b, we can now derive the following expression for evolutionary change:

$$\Delta p = (rb - c) \cdot \mathrm{Var}(p)/\overline{w} \tag{5.7}$$

where $\overline{w}$ is average population fitness (see Appendix 5.1). Equation (5.7) is a generalized version of Hamilton's rule, which holds true irrespective of whether payoffs are additive. It tells us that when $0 < p < 1$, the A type will increase in frequency in the next generation if and only if $rb > c$. Importantly, this version of the rule holds true whether payoffs are additive or not. It constitutes a quite general way of partitioning the total evolutionary change into direct and indirect components.

The quantity $(rb - c)$, whose sign determines whether altruism spreads, can be computed from equations (5.5) and (5.6). After simplifying, this yields:

$$rb - c = (rB - C) + d[r + p(1 - r)] \tag{5.8}$$

Note that $(rb - c)$ is a function of p, so satisfaction of $rb > c$ in one generation does not imply its satisfaction in the next. Selection is thus frequency-dependent, and neither type will necessarily spread to fixation. A polymorphic equilibrium will obtain when $p = [C - r(B + D)]/D[1 - r]$; the stability of this equilibrium depends on the sign of D. The full evolutionary dynamics are summarized in Table 5.9; see Appendix 5.1 for proof.

The generalized form of Hamilton's rule embodied in equation (5.7) raises interesting interpretive questions. Some have argued that the rule in this form has little explanatory value, while others have seen the generality of the rule as an advantage, a proof that inclusive fitness theory does not rely on restrictive assumptions. This debate is explored in section 5.5. For now, our question is this. Given that equation (5.7) is true, and given the resulting evolutionary dynamics, can agential thinking (type 1) be salvaged? Will evolution lead an organism to behave like a rational agent trying to maximize a utility function, and if so, what is it?

Importantly, the answer to this question cannot simply be read off equation (5.7). In the additive case there was a simple link between Hamilton's rule and a utility function with the desired property: $rB - C > 0$ was the condition for the A type to spread, and $(rB - C)$ the utility difference between playing A and S. One might hope

Table 5.9. Evolutionary dynamics, non-additive case

Case 1: $r < 1, D > 0$	
(i) $rB - C + rD \geq 0$	*A* evolves to fixation
(ii) $rB - C + D \leq 0$	*S* evolves to fixation
(iii) $rB - C + D > 0 > rB - C + rD$	unstable polymorphism at $p = [C - r(B + D)]/D[1 - r]$
Case 2: $r < 1, D < 0$	
(i) $rB - C + D \geq 0$	*A* evolves to fixation
(ii) $rB - C + rD \leq 0$	*S* evolves to fixation
(iii) $rB - C + D < 0 < rB - C + rD$	stable polymorphism at $p = [C - r(B + D)]/D[1 - r]$
Case 3: $r = 1$	
(i) $B - C + D > 0$	*A* evolves to fixation
(ii) $B - C + D < 0$	*S* evolves to fixation
(iii) $B - C + D = 0$	no evolutionary change

to extrapolate this to the non-additive case by simply replacing $(rB - C)$ with $(rb - c)$ in the utility function in Table 5.4. However, since b and c are functions of p, they cannot meaningfully feature in the agent's utility function. For the point of organism-as-rational-agent idea is to find a link between the outcome of evolution and a goal that organisms behave as if they are trying to achieve, that is, a quantity they try to maximize. For this to make sense, it is essential that the goal does not keep changing as the population evolves.

5.4.3 Rational agent analysis

To address this question, we again transpose the evolutionary model to a rational choice context and study the Nash equilibria of the resulting game. Suppose first that the utility function is given by the simplified IF payoff transformation, that is, personal payoff plus r times partner payoff. This yields the payoffs in Table 5.10.

The Nash equilibria are then as follows:

(A, A) is a Nash equilibrium if and only if $rB - C + D(r + 1) \geq 0$

(S, S) is a Nash equilibrium if and only if $rB - C \leq 0$

(π_A, π_A) is a mixed strategy Nash equilibrium where $\pi_A = (C - rB)/D(1 + r)$, as long as $0 < \pi_A < 1$.

It follows that, unlike in the additive case, the organism-as-rational agent concept will not work when utility is defined as (simplified) inclusive fitness. The condition for (A, A) to be a Nash equilibrium is not identical to the condition for A to evolve to fixation; similarly for S. Furthermore, the condition for there to be a mixed-strategy Nash equilibrium is not the same as the condition for there to be a polymorphism. So it is not true that at evolutionary equilibrium, organisms will behave as if trying to maximize their inclusive fitness.

Table 5.10. Non-additive case, simplified IF payoffs

		Partner	
		A	S
Actor	A	$(B - C + D)(r + 1)$	$-C + rB$
	S	$B - rC$	0

Table 5.11. Non-additive case, Grafen 1979 payoffs

		Partner	
		A	S
Actor	A	$(B - C + D)$	$-C + rB + rD$
	S	$(1 - r)B$	0

Can we find a utility function that allows agential thinking to be salvaged? The answer is yes. The Grafen 1979 payoff matrix, which as we saw is derived from the personal payoff matrix by the transformation $U(i,j) = rV(i,i) + (1-r)V(i,j)$, does the trick. This yields the payoffs in Table 5.11.

The Nash equilibria are then as follows:

(A, A) is a Nash equilibrium if and only if $rB - C + D \geq 0$

(S, S) is a Nash equilibrium if and only if $rB - C + rD \leq 0$

(π_A, π_A) is a mixed strategy Nash equilibrium, where $\pi_A = (C - r(B + D))/D(1 - r)$, as long as $0 < \pi_A < 1$.

This yields the desired outcome. In particular, if (*A*, *A*) is the only pure-strategy Nash equilibrium, then *A* evolves to fixation; if (*S*, *S*) is the only pure-strategy equilibrium, then *S* evolves to fixation. If there is a mixed-strategy Nash equilibrium but no pure strategy equilibria, the population evolves to a stable polymorphism; if there is a mixed-strategy Nash equilibrium and both (*A*, *A*) and (*S*, *S*) are pure-strategy equilibria, then there is an unstable polymorphism; in both cases, the weights on *A* and *S* in the mixed-strategy Nash equilibrium equal the proportions of *A* and *S* in the polymorphism. Thus there is a tight correspondence between the Nash equilibria and the evolutionary dynamics, summarized in Table 5.12.

The upshot is that with non-additive payoffs, the organism-as-rational-agent heuristic will work as long as the utility function is defined as Grafen 1979 payoff, rather than inclusive fitness payoff. Again any positive linear transformation of the Grafen 1979 payoff matrix, or any local shift, will also preserve the correspondences above. Note that, unlike in the additive case, the Grafen 1979 payoff matrix (Table 5.11) is *not* a local shift of the (simplified) IF payoff matrix (Table 5.10). This is why the heuristic fails if utility is defined as inclusive fitness in the non-additive case.

Table 5.12. Evolution–rationality link, utility = Grafen 1979 payoff

(A, A) is only pure N.E.	$\Longrightarrow$	A evolves to fixation
(S, S) is only pure N.E.	$\Longrightarrow$	S evolves to fixation
(π_A, π_A) is only N.E.	$\Longleftrightarrow$	stable polymorphism at $p = \pi_A$
(π_A, π_A), (A, A), (S, S) all N.E.	$\Longleftrightarrow$	unstable polymorphism at $p = \pi_A$

Note: $\pi_A = (c - r(b + d)/d(1 - r))$

5.4.4 Discussion

Let us briefly recap. In our simple pairwise interaction model with additive payoffs, Hamilton's rule holds true and natural selection pushes the population to an equilibrium in which individuals maximize their inclusive fitness. So agential thinking (type 1) works fine, with utility defined as inclusive fitness (simplified or original), just as Hamilton and his modern supporters argue.

With non-additive payoffs, matters are different. Although a version of Hamilton's rule can be applied in this case, by suitably defining the cost and benefit terms, we find that at evolutionary equilibrium, organisms do not behave as if trying to maximize their (simplified) inclusive fitness. However, there is a somewhat similar quantity—Grafen 1979 payoff—that they do behave as if they are trying to maximize.

It is an open question whether our positive result—maximization of Grafen 1979 payoff—extends to more complicated models of social evolution, for example, that incorporate kin competition, multiple social partners, or class structure, or to more realistic genetic architectures than haploid inheritance. There is no guarantee that it does, as such models typically lead to more complicated evolutionary dynamics than those assumed here. A valid maximization argument must always deduce the quantity being maximized, if any, from the underlying dynamics (Mylius and Diekmann 1995).

Also, we have assumed that the coefficient of relatedness, r, remains constant as the population evolves. Without this assumption, it makes little sense to allow the utility function to depend on r, as this would be tantamount to positing a changing goal, so would make nonsense of the idea of treating organisms as akin to rational agents trying to maximize some quantity through their choice of action. In some inclusive fitness models, r is in fact a dynamic variable rather than a constant, so it cannot be assumed that our positive result, or one like it, can be derived for these models.[14]

Our negative result, that maximization of inclusive fitness only holds with additive payoffs, partly supports the claims made by opponents of inclusive fitness theory. In a recent critique, Allen *et al.* (2013) write: 'evolution does not, in general, lead to the maximization of inclusive fitness or any other quantity', noting that arguments to the contrary assume payoff additivity (p. 20138). Here we have understood maximization to include best-response, so the presence of frequency-dependence does not

[14] For an example of such a model, see van Baalen and Rand (1998).

automatically preclude a maximization principle from holding; and we have employed the 'simplified' definition of an individual's inclusive fitness that is well-defined in the non-additive case, unlike the original definition. But even so, individuals do not maximize their inclusive fitness at equilibrium.

The key logical point is that although a version of Hamilton's rule is indeed a fully general evolutionary truth, as Gardner *et al.* (2011) stress, nothing about individual maximization can be deduced directly from this generalized form of the rule, contrary to what is sometimes implied. In particular, although the generalized Hamilton's rule yields the $rb > c$ condition for the social trait to spread, one cannot conclude that individuals behave as if trying to maximize the quantity $(rb - c)$; they cannot do this, as this quantity is frequency-dependent. (This is the point missed by Gardner *et al.* (2011), who suggest, apropos the generalized Hamilton's rule, 'we can imagine the individual adjusting her inclusive fitness . . . by altering her behaviour' (pp. 1039–40).) Whether an individual maximization principle holds, and if so what the quantity being maximized is, needs to be analysed on a case-by-case basis.

Proponents of the inclusive fitness concept argue that it offers behavioural ecologists a way to interpret social behaviour in the field. Our analysis suggests that even in simple cases this may not be possible. If an observed social behaviour fails to increase inclusive fitness, defined as personal payoff plus r times partner's payoff, the behaviour may nonetheless be adaptive and the population at an evolutionary equilibrium. Moreover, the quantity we have called 'Grafen 1979 payoff' will serve the needs of a biologist seeking to identify the goal of evolved behaviour in a broader range of cases than will inclusive fitness itself.

5.5 Causality and Switching

A second dimension of the recent controversy concerns the biological significance of the generalized Hamilton's rule. We saw that in Hamilton's original work, the costs and benefits of a social action are described in explicitly causal terms; his aim was to decompose natural selection into distinct causal components (Frank 2013). However, a persistent complaint against the generalized Hamilton's rule, in which the costs and benefits are partial regression coefficients, is that its terms lack a natural causal meaning (Allen and Nowak 2015). Generality has been purchased at the expense of meaningfulness, the critics argue.[15]

To address this issue, we continue with the pairwise interaction model of section 5.3, with non-additive payoffs (Table 5.8). Let us focus on the cost of the social action. (A similar analysis applies to the benefit term.) How should we measure this cost? A natural way is to consider a hypothetical experimental intervention. We randomly pick an organism that does not perform the social action, that is, an *S* type, and switch

[15] The material in this section is based on Okasha and Martens (2016b).

it to A while leaving everything else unchanged; and we consider the effect on the actor's personal payoff. Doing an experimental intervention of this sort is the standard way to assess causality in science, and is often taken to define the causal relation (e.g. Pearl 2000). In the context of social evolution models, such 'switching' has often been discussed as a way of assessing the effect of a social action.[16] Switching also relates to agential thinking; for a rational agent, deliberating about what to do is precisely concerned with what the effect on its payoff would be if it were to switch from one action to another.

What is the effect of such a switch? If the chosen S type (the actor) has an A partner, then switching will increase the actor's payoff by $(-C+D)$. If the actor has an S partner, then switching will increase the actor's payoff by $-C$. The expected effect of the $S \to A$ switch on the actor's payoff is therefore a weighted average of $(-C+D)$ and $-C$, with the weights given by the conditional probability that the chosen actor has a partner of each type. Therefore we have:

Expected causal effect of $S \to A$ switch on actor's payoff

$$\begin{aligned} &= (-C+D) \cdot \Pr(\text{partner is } A \mid \text{actor is } S) + (-C) \cdot \Pr(\text{partner is } S \mid \text{actor is } S) \\ &= (-C+D) \cdot p(1-r) + (-C) \cdot [1 - p(1-r)] \qquad (5.9) \\ &= -C + D \cdot p(1-r) \end{aligned}$$

The quantity in (5.9) is called the expected *causal* effect because it describes the result of an experimental intervention, so is not simply a population statistic like a regression coefficient. The experiment consists in switching a randomly chosen S type into an A while holding fixed its partner. The outcome of the experiment, that is, the effect on the actor's payoff, is a random variable which can take two values, $(-C+D)$ or $(-C)$; the expected value of this random variable is equation (5.9).

Another way to interpret this expected value is to consider a cohort of S types drawn from the population. Some members of the cohort will be partnered with Ss, others with As. If the cohort is representative of the population, the proportions partnered with an S and an A will be Pr(partner is A | actor is S) and Pr(partner is S | actor is S), respectively. Suppose that all members of the cohort then switch from S to A while their partners are held fixed. This will cause a per-capita change in personal payoff equal to the expected effect in equation (5.9).

Note that the expected causal effect of an $S \to A$ switch on the actor's payoff is not equal but opposite in sign to the expected causal effect of the reverse $A \to S$ switch (unless $D=0$ or $r=0$). This is because a randomly chosen A faces different probabilities of having a partner of each type than does a randomly chosen S.

[16] See Nunney (1985), Karlin and Matessi (1983), Allen and Nowak (2015), and Peña *et al.* (2015).

Now compare equation (5.9) with the $-c$ term in the generalized Hamilton's rule (5.5), which for convenience is repeated here:

$$-c = (-C) + (D) \cdot [r + p(1 - r)]/[1 + r] \tag{5.5}$$

The $-c$ term is also a weighted average of $(-C + D)$ and $(-C)$, *but with different weights*. The two terms are identical in exactly three cases: (i) $D = 0$; (ii) $r = 0$; (iii) $p = 1/(1 - r)$. Case (i) is where payoffs are additive; case (ii) is where pairs are formed at random; case (iii) is only possible if $r \leq 0$, that is, negative assortment, and for fixed r the equality in question will obtain for only one value of p. Thus with non-additive payoffs, the $-c$ term of Hamilton's rule and the expected causal effect of an $S \rightarrow A$ switch on actor's payoff will almost always differ in magnitude, and may differ in sign.

It is useful to express $-c$ in a form that permits direct comparison with the expected causal effect:

$$-c = (-C + D)\frac{\Pr(\text{partner is } A \mid \text{actor is } A)}{k} + (-C)\frac{\Pr(\text{partner is } S \mid \text{actor is } S)}{k} \tag{5.10}$$

where $k = [\Pr(\text{partner is } A \mid \text{actor is } A) + \Pr(\text{partner is } S \mid \text{actor is } S)]$

In equation (5.10), the weights on $(-C + D)$ and $(-C)$ are proportional to $\Pr(\text{partner is } A \mid \text{actor is } A)$ and $\Pr(\text{partner is } S \mid \text{actor is } S)$, respectively. (These two probabilities do not sum to one, hence the normalizing term k in the denominator.) When $-c$ is written this way, its oddity as a measure of the causal effect of the social action becomes apparent. The components in the sum, $(-C + D)$ and $(-C)$, have an obvious causal meaning; they are the changes to an actor's personal payoff caused by an $S \rightarrow A$ switch, depending on its partner type. But the weights on these two changes do not equal the proportion of S types in the population with a partner of each sort; so $-c$ is not the per-capita change in personal payoff that would result if a representative cohort of S types were switched to A.

The discrepancy between $-c$ and the expected causal effect is due to non-additivity. In general, a partial regression coefficient only corresponds to the expected effect of an experimental intervention of the sort described here—in which one independent variable is increased by a unit while the other(s) are held fixed—if the linear model describes the true relation between the variables (Pearl 2001, Gelman and Hill 2007). In the current case, where a linear model has been fitted to a non-linear system, $-c$ does not equal the expected effect on actor's payoff resulting from experimentally switching a randomly chosen S type to A while holding its partner fixed.

This analysis supports the view that the $-c$ term of the generalized Hamilton's rule, while useful for describing evolutionary change, lacks a natural causal interpretation. Experimental determination of the cost of the social action, via the experiment described above, will not agree with the cost as measured by the partial regression coefficient. Parallel remarks apply to the relation between the b term of the generalized Hamilton's rule and the expected effect of an $S \to A$ switch on partner payoff.

5.5.1 *Fisher to the rescue?*

One way of salvaging the causal meaning of the generalized Hamilton's rule is to draw on an argument made by R. A. Fisher in a different context, recently revisited by Lee and Chow (2013). To derive his fundamental theorem of natural selection, Fisher (1930) introduced a notion he called 'the average effect of a gene substitution' on a quantitative character of interest (which could be fitness). This was intended as a measure of the effect, on average in a population, of a given gene being substituted for one of its alleles (e.g. by mutation), and was defined by Fisher in terms of the linear regression of an organism's character value on the number of copies of the gene that it has. We previously encountered this notion in section 3.3.

Fisher (1941) focuses on the average effect of a gene substitution in a one-locus two-allele Mendelian model with dominance. There are three genotypes AA, AB, and BB in a population with frequencies of P, $2Q$, and R; character values are w_{AA}, w_{AB}, and w_{BB} (Table 5.13). Random mating is not assumed, so genotypes need not be in Hardy-Weinberg proportions. The effect of a $B \to A$ substitution depends on whether the substituted gene is in a homozygote or heterozygote, that is, on whether the change is from $BB \to AB$ or from $AB \to AA$. The former substitution changes an individual's character by $[w_{AB} - w_{BB}]$, the latter by $[w_{AA} - w_{AB}]$. The average effect of the gene substitution is a weighted average of these two quantities.

Though Fisher's average effect is defined statistically, via the linear model, he also gives it a causal interpretation. Fisher (1941) *appears* to claim that the average effect of a $B \to A$ substitution equals the average change in character if a randomly picked B allele were experimentally changed into an A while everything else is held constant. Intuitively this interpretation is suspect, given that a linear model has been fitted to a non-linear system; and indeed Falconer (1985) argued that Fisher's interpretation was incorrect, by computing the expected effect of a hypothetical $B \to A$ mutation and showing that it does not, in general, equal Fisher's average effect of a gene substitution. However, more recently Lee and Chow (2013), building on Edwards (2002), have unpacked Fisher's curious logic and argued that, *in a sense*, the average effect can be imbued with a causal meaning even under dominance.

Table 5.13. One-locus two-allele model

Genotype	Frequency	Character value
AA	P	w_{AA}
AB	$2Q$	w_{AB}
BB	R	w_{BB}

The issue is formally very similar to our social evolution problem (and we have changed Fisher's notation to emphasize this). Our three pair-types correspond to the three genotypes in Fisher's model; personal payoff corresponds to character value; and an $S \to A$ switch corresponds to an $B \to A$ substitution. In both cases the effect of the switch (or substitution) is context-dependent, due to the non-linearity; and in both cases we have a measure of the average effect of the switch (or substitution) defined by fitting a linear model. Thus the $-c$ term in the generalized Hamilton's rule corresponds to Fisher's average effect of a gene substitution.

The key point is that Fisher was concerned with the effect of a gene substitution 'in the population as actually constituted', as Lee and Chow (2013) emphasize. Fisher understood the 'constitution' of the population to include the rules by which the genes are combined into genotypes, that is, the mating pattern, as discussed in section 3.3. Substituting a number of *B* genes by *A* genes might break the rules of combination in the population, that is, require an implicit change in the mating pattern, so the actual effect of such an intervention might not correspond to the effect with which Fisher was concerned.

Fisher (1941) introduces a particular measure λ of the deviation from random mating in the population, defined by $\lambda = Q^2/PR$. With random mating, that is, Hardy-Weinberg proportions, $\lambda = 1$; with assortative mating, $\lambda > 1$. Constancy of the mating pattern, in Fisher's discussion, means constancy of λ; this concept was discussed in section 3.3.

Now consider the effect of a $B \to A$ substitution on λ. If the substitution is of the $BB \to AB$ sort then it will increase λ, while if it is of the $AB \to AA$ sort then it will reduce λ. Suppose we pick a cohort of *B* genes and substitute them with *A* genes. In order for this intervention to leave λ unchanged, the *B* genes in the cohort must come from *BB* and *AB* individuals in specific proportions; so the cohort must be carefully chosen. What Fisher (1930) shows is that the per-capita change in character value, in the cohort, then equals the average effect of the gene substitution as defined by the linear model.

An equivalent way to formulate Fisher's result is this. The average effect of a $B \to A$ substitution, as defined by the linear model, is a weighted average of $[w_{AB} - w_{BB}]$ and $[w_{AA} - w_{AB}]$, which are the changes in individual character value that result from a $BB \to AB$ and an $AB \to AA$ mutation, respectively. The weights are defined by the proportions of *BB* and *AB* individuals in a cohort which is such that, when all the *B* genes in the cohort are switched to *A*, λ is unchanged. Importantly, the proportions of *BB* and *AB* individuals in such a cohort will not in general equal their population-wide proportions.

This goes some way towards reconciling Fisher's statistical definition of the average effect with the hypothetical experimental intervention he describes. Falconer (1985) was right that Fisher's average effect is not equal to the expected character change of a $B \to A$ substitution if the *B* gene is picked at random from the whole population. However, if the *B* gene is picked at random from a cohort that meets Fisher's 'constant λ' condition, then the resulting expected change is equal to the average effect of the gene substitution as defined by Fisher. Applied to social evolution, Fisher's argument provides a possible way of salvaging the causal meaning of the generalized Hamilton's rule.

5.5.2 *Application to social evolution*

Returning to the pairwise interaction model, consider the following experiment. We pick any cohort of S types from the population and switch them to A. As a result of this intervention, P, Q, and R change by $\mathrm{d}P$, $\mathrm{d}Q$, and $\mathrm{d}R$, respectively, where $\mathrm{d}P + \mathrm{d}Q + \mathrm{d}R = 0$; therefore p changes by $\mathrm{d}p = \mathrm{d}P + \mathrm{d}Q$.

When an S in an AS pair is switched to A, its payoff increases by $(-C + D)$; when an S in an SS pair is switched to A, its payoff increases by $(-C)$. The ratio of the two types of switches is $\mathrm{d}P : -\mathrm{d}R$. Therefore, the per-capita change in an actor's payoff caused by the experimental intervention equals:

$$\frac{dP(-C + D) - dR(-C)}{dP - dR} \tag{5.11}$$

Expression (5.11) holds true irrespective of how the cohort of S types is chosen. If the cohort is chosen at random, that is, contains S types with A and S partners in identical proportions to those in the global population of S types, then $\mathrm{d}P : -\mathrm{d}R = Q : R$, and equation (5.11) is then equal to the expected causal effect of an $S \rightarrow A$ switch on the actor's payoff, as defined by equation (5.9).

Now consider the $-c$ term in the generalized Hamilton's rule, given in equation (5.5). Following Fisher's logic, we equate $-c$ with the per-capita change in an actor's payoff caused by the experimental intervention (5.11), and extract a constraint on $\mathrm{d}P$, $\mathrm{d}Q$, and $\mathrm{d}R$. Equating (5.5) and (5.11) gives:

$$\frac{dP(-C + D) - dR(-C)}{dP - dR} = -C + \frac{r + p(1 - r)}{1 + r} \cdot D$$

We then make the following substitutions:

$$\begin{aligned} -\mathrm{d}R &= \mathrm{d}P + 2\mathrm{d}Q \\ p &= P + Q \\ P &= rp + p^2(1 - r) \\ r &= [P - (P + Q)^2]/[(P + Q)(R + Q)] \end{aligned}$$

After simplifying, this gives:

$$\mathrm{d}P(QR - QP) = 2\mathrm{d}Q(PR + PQ)$$

Dividing across by PQR and further simplifying yields:

$$(\mathrm{d}P/P) + (\mathrm{d}R/R) = 2(\mathrm{d}Q/Q) \tag{5.12}$$

which is exactly the constraint found in Fisher (1941).

Following Fisher, we then solve equation (5.12) by taking the infinite limit, that is, letting dP, dQ, and dR become arbitrarily small.[17] This gives:

$$Q^2/PR = \text{constant} = \lambda \tag{5.13}$$

This is precisely Fisher's 'constant λ' condition, and shows the close link between our social evolution problem and his population-genetic problem. The meaning of equation (5.13) is worth rehearsing. A cohort of S types, some in AS and some in SS pairs, was chosen from the population and experimentally switched to A. We then asked the question: under what condition is the per-capita change in an actor's payoff that results from the experimental intervention equal to $-c$? The answer is given by equation (5.13): the cohort must be chosen in such a way that the intervention leaves the ratio Q^2/PR unchanged. This then determines the proportions of S types in the cohort with A and S partners, respectively.

It follows that the $-c$ term of Hamilton's rule does have a quasi-causal meaning, even with non-additive payoffs. As we know, $-c$ is not the expected change in an actor's payoff if a randomly chosen S type is switched to A. However, if the S to be switched is chosen not at random from the population, but rather at random from any cohort of Ss satisfying the 'constant λ' condition, then the resulting expected change in payoff is equal to $-c$. If we regard the assortment pattern as an 'environmental' parameter, quantified by λ, and wish to measure the causal effect of an $S \to A$ switch on an actor's payoff in a constant environment, then $-c$ is arguably the correct measure. A parallel analysis applies to the b term.

5.5.3 *Constant r versus constant λ*

This Fisherian defence of the causal meaning of the generalized Hamilton's rule rests on three premises: first, that in assessing the causal effect of an $S \to A$ switch the environment should be held constant; second, that the assortment pattern is part of the environment; and third, that λ is the appropriate measure of assortment.

The third premise is the hardest to defend. For an alternative measure of assortment is simply r itself, the coefficient of relatedness between social partners. Recall the respective definitions of r and λ in terms of P, Q, and R:

$$r = \frac{P - (P+Q)^2}{(P+Q)(R+Q)}$$

$$\lambda = Q^2/PR$$

Algebraic manipulation shows that r and λ are related as follows:

$$\lambda = \frac{Q(1-r)}{Q(1-r)+r} \tag{5.14}$$

[17] To derive equation (5.13) from equation (5.12), simply integrate both sides and combine the constants of integration, noting that $\int x^{-1}dx = \ln|x| + c$.

Equation (5.14) shows that r and λ both quantify the deviation from random assortment. Note that r ranges from -1 to $+1$, while λ ranges from 0 to $+\infty$. With random assortment, $r = 0$ and $\lambda = 1$; with perfect assortment, $r = 1$ and $\lambda = 0$. However, constancy of λ across generations does not imply constancy of r, nor vice-versa. The following numerical example illustrates this point.

Example

In generation 1, $p = \frac{1}{2}, r = \frac{1}{2}$.
Therefore, $P = \frac{3}{8}, 2Q = \frac{1}{4}, R = \frac{3}{8}$ and $\lambda = \frac{1}{9}$.
Evolution occurs, leading p to increase to $\frac{3}{4}$.

Case (i): r stays constant

So in generation 2, $p = \frac{3}{4}, r = \frac{1}{2}$.
Therefore, $P = \frac{21}{32}, 2Q = \frac{6}{32}, R = \frac{5}{32}$ and $\lambda = \frac{3}{35}$.
So r has stayed constant while λ has decreased.

Case (ii): λ stays constant

So in generation 2, $p = \frac{3}{4}, \lambda = \frac{1}{9}$.
Therefore, $P \approx 0.65, 2Q \approx 0.20, R \approx 0.15$.
This gives $r \approx 0.47$. So λ has stayed constant while r has decreased.

Fisher (1941) offers no independent argument for why λ is the 'correct' measure of deviation from random mating in his population-genetic model, nor therefore for why environmental constancy should mean constancy of λ. Rather, he simply shows that constancy of λ is implied if we equate the average effect of a gene substitution, as defined by the linear model, with the per-capita effect caused by an experimental gene substitution.

The same applies to our social evolution model. If environmental constancy were defined as constancy of r rather than λ, this would imply different weights on the two sorts of $S \rightarrow A$ switch; and if we computed the expected effect of an $S \rightarrow A$ switch on an actor's payoff using these weights, the result would not equal $-c$. In the absence of an independent reason to hold λ rather than r fixed, the Fisherian defence of the causal meaning of Hamilton's rule, above, cannot be considered logically watertight.

5.5.4 Discussion

With additive payoffs, the cost and benefit terms of Hamilton's rule have a clear causal meaning: they equal the amount by which an actor would increase its own and its social partner's payoff, respectively, by performing the social action, that is, switching from S to A. With non-additivity the situation is different, as the effect of the switch depends on context, that is, partner type. Since the expected effect of an $S \rightarrow A$ switch on an actor's payoff does not equal the $-c$ term, and similarly for b, the causal meaning of the generalized Hamilton's rule, derived from the regression method, is called into question.

A possible way of rescuing the causal meaning of the $-c$ and b terms involves adapting Fisher's idea that his average effect of a gene substitution, in a non-additive genetic system, corresponds to the expected result of an experimental intervention in a 'constant environment', so does have a causal meaning. An analogous argument applies exactly to our social evolution model with synergistic payoffs. However, like Fisher's original, the argument has an intrinsic limitation in that it relies on a particular way of defining environmental constancy that lacks independent justification.

The upshot, therefore, is that the defenders and critics of the generalized Hamilton's rule are both partly right. When $-c$ and b are defined via the regression method, they do not correspond to the cost and benefit of the social action as measured by a standard experimental determination. However, it does not follow that the generalized Hamilton's rule is devoid of all causal meaning, nor that the definitions of $-c$ and b are simply arbitrary constructs designed to preserve the truth of the rule, as is sometimes alleged. For as Fisher shows, his average effect can be interpreted as the expected outcome of an experimental intervention of a very particular sort, and precisely the same is true of the components of Hamilton's rule.

5.6 Conclusion

Inclusive fitness and kin selection are widely used tools for studying social evolution, but controversy over their status persists. Here we have focused on two issues in particular: inclusive fitness maximization, and the causal meaning of the generalized Hamilton's rule. The former issue is intimately related to agential thinking; part of Hamilton's motivation was to find a quantity which organisms behave as if trying to maximize in social contexts, and this motivation accounts for much of the popularity of the inclusive fitness concept today. The second issue also relates to agential thinking, obliquely, since causality is an essential aspect of rational deliberation.

Our principal finding is that although the generalized Hamilton's rule is indeed a fully general truth about evolution, and does have a causal meaning of sorts, this in itself does not validate the idea that evolution will lead individuals to behave as if their goal were to maximize inclusive fitness. The latter idea appears to have been oversold, as its theoretical basis is slighter than some of its advocates acknowledge, requiring restrictive assumptions. In this case, it seems that agential thinking (type 1) may be partly to blame. The intuitive appeal of the organism-as-rational-agent concept, I suggest, has helped foster the belief that inclusive fitness maximization is more generally applicable than it really is.

However, I do not think we should jettison the idea of inclusive fitness as the goal of social behaviour altogether, for two reasons. First, if social interactions have weak effects on fitness, then the additivity assumption, and thus inclusive fitness maximization, may obtain as a reasonable approximation. Second, the real proving ground of the idea is in the empirical domain. If inclusive fitness allows behavioural ecologists to explain and predict what they see, the fact that it lacks generality as a theoretical principle may not matter so much. This is an instance of the moral stressed in chapter 4: to the extent that agential thinking (type 1) is justified, it is empirical facts, and not some overarching principle about natural selection, that will ultimately justify it.

Appendix 5.1

Additive payoffs

(i) By the covariance formula of Price (1970), we have:

$$\Delta p = Cov(w_i, p_i)/\overline{w} = p(w_A - \overline{w})/\overline{w}$$

where w_i is fitness of ith individual, and w_A is average fitness of A type.

From Table 5.1 and the definition of r:

$$w_A = 1 + b[(r + p(1 - r)] - c$$
$$\overline{w} = 1 + p[b - c]$$

Substituting into the expression for Δp and simplifying:

$$\Delta p = (rb - c) \cdot p(1 - p)/\overline{w}$$
$$= (rb - c) \cdot Var(p)/\overline{w}$$

which is equation (5.3).

(ii) When utility = inclusive fitness payoff (Table 5.4), then by the definition of Nash Equilibrium (N.E.):

$$(A, A) \text{ is an N.E.} \iff (b - c)(r + 1) \geq b - rc \iff rb \geq c$$
$$(S, S) \text{ is an N.E.} \iff 0 \geq -c + rb \iff rb \geq c$$

Therefore (A, A) and (B, B) are both N.E. $\iff rb = c$.

Let π_A be an arbitrary mixed strategy that plays A with probability π_A. Then by the definition of mixed strategy payoff:

$$(\pi_A, \pi_A) \text{ is a N.E.} \iff U(A, \pi_A) = U(S, \pi_A)$$

where $U(A, \pi_A)$ is the expected payoff to strategy A played against π_A etc.

By Table 5.4:

$$U(A, \pi_A) = (b - c)(r + 1)\pi_A + (rb - c)(1 - \pi_A)$$
$$U(S, \pi_A) = \pi_A(b - rc)$$

Therefore $U(A, \pi_A) = U(S, \pi_A) \iff rb = c \iff (\pi_A, \pi_A)$ is an N.E.

This establishes the correspondences in Table 5.6.

Non-additive payoffs

(i) By the covariance formula of Price (1970), we have:

$$\Delta p = Cov(w_i, p_i)/\overline{w}$$

From equation (5.4) we have:

$$w_i = \alpha + -Cp_i + Bp'_i + e_i$$

Substituting into the expression for Δp:

$$\begin{aligned}\Delta p &= [-C \cdot Cov(p_i, p_i) + B \cdot Cov(p'_i, p_i)]/\overline{w} \\ &= [-C \cdot Var(p) + B \cdot b_{p'p} \cdot Var(p)]/\overline{w} \\ &= (-C + rB) \cdot Var(p)/\overline{w}\end{aligned}$$

which is equation (5.7).

(ii) Equation (5.8) tells us that $rB - C = (rb - c) + d[r + p(1 - r)]$.
Therefore, A evolves to fixation from any initial frequency:

$$\Longleftrightarrow (rb - c) + d[r + p(1 - r)] > 0, \forall p \in (0, 1)$$

Similarly, S evolves to fixation from any initial frequency:

$$\Longleftrightarrow (rb - c) + d[r + p(1 - r)] < 0, \forall p \in (0, 1)$$

Polymorphic equilibrium obtains when $rB - C = 0$. Solving for p gives:

$$p^* = [c - r(b + d)]/d[1 - r]$$

If $d > 0$, then $\Delta p < 0$ when $p < p^*$ and $\Delta p > 0$ when $p > p^*$, so the equilibrium is unstable. If $d < 0$ the equilibrium is stable.

Suppose that $r < 1$ and $d > 0$ (case 1).
Then, $(rb - c) + d[r + p(1 - r)] \to (rb - c + rd)$ as $p \to 0$
and $(rb - c) + d[r + p(1 - r)] \to (rb - c + d)$ as $p \to 1$.
Therefore, A evolves to fixation $\Longleftrightarrow (rb - c) + rd \geq 0$.
Similarly, S evolves to fixation $\Longleftrightarrow (rb - c) + d \leq 0$.
When $rb - c + d > 0 > rb - c + rd$, then $0 < [c - r(b + d)]/d[1 - r] < 1$,
so an unstable polymorphism evolves.

Suppose that $r < 1$ and $d < 0$ (case 2).
By identical reasoning to case 1:
A evolves to fixation $\Longleftrightarrow (rb - c) + d \geq 0$.
S evolves to fixation $\Longleftrightarrow (rb - c) + rd \leq 0$.
When $rb - c + d < 0 < rb - c + rd$, then $0 < [c - r(b + d)]/d[1 - r] < 1$,
so a stable polymorphism evolves.

Suppose that $r = 1$ (case 3).
Then, $rB - C = b - c + d$. Therefore:
A evolves to fixation $\Longleftrightarrow b - c + d > 0$.
S evolves to fixation $\Longleftrightarrow b - c + d < 0$.
No evolutionary change occurs $\Longleftrightarrow b - c + d = 0$.
This establishes the evolutionary dynamics in Table 5.9.

(iii) When utility = inclusive fitness payoff (Table 5.10), then by the definition of Nash equilibrium:

$$(A, A) \text{ is a N.E.} \iff (b - c + d)(r + 1) \geq b - rc \iff rb - c + d(r + 1) \geq 0$$
$$(S, S) \text{ is a N.E.} \iff rb - c \leq 0$$

Let π_A be an arbitrary mixed strategy. Then, (π_A, π_A) is an N.E. if and only if $U(A, \pi_A) = U(S, \pi_A)$. From Table 5.10:

$$U(A, \pi_A) = \pi_A(b - c + d)(r + 1) + (1 - \pi_A)(rb - c)$$
$$U(S, \pi_A) = \pi_A(b - rc)$$

Therefore, (π_A, π_A) is an N.E. $\iff \pi_A = (c - rb)/d(1 + r)$.

(iv) When utility = Grafen 1979 payoff (Table 5.11), then by the definition of Nash equilibrium:

$$(A, A) \text{ is a N.E.} \iff (b - c + d) \geq (1 - r)b \iff rb - c + d \geq 0$$
$$(S, S) \text{ is a N.E.} \iff rb - c + rd \leq 0$$

Let π_A be an arbitrary mixed strategy. Then, (π_A, π_A) is an N.E. if and only if $U(A, \pi_A) = U(S, \pi_A)$. From Table 5.11:

$$U(A, \pi_A) = \pi_A(b - c + d) + (1 - \pi_A)(rb + rd - c)$$
$$U(S, \pi_A) = \pi_A(1 - r)b$$

Therefore, (π_A, π_A) is a N.E. $\iff \pi_A = (c - r(b + d))/d(1 - r)$.

If (A, A) is sole pure-strategy N.E.

$$\iff rb - c + d \geq 0 \text{ and } rb - c + rd > 0$$
$$\implies (rb - c) + d[r + p(1 - r)] > 0, \forall p \in (0, 1)$$
$$\iff A \text{ spreads to fixation.}$$

Similarly, (S, S) is sole pure-strategy N.E. $\implies S$ spreads to fixation.

Consider mixed strategy (π_A, π_A), where $\pi_A = (c - r(b + d))/d(1 - r)$. Then, (π_A, π_A) is the only N.E.

$$\iff rb - c + rd > 0 > rb - c + d \text{ and } 0 < (c - r(b + d))/d(1 - r) < 1$$
$$\iff rb - c + rd > 0 > rb - c + d \text{ and } d < 0 \text{ and } r < 1$$
$$\iff \text{stable polymorphism obtains at } p = \pi_A.$$

Also, (π_A, π_A), (A, A), and (S, S) are all N.E.

$$\iff rb - c + d \geq 0 \geq rb - c + rd \text{ and } 0 < (c - r(b + d))/d(1 - r) < 1$$
$$\iff rb - c + d > 0 > rb - c + rd \text{ and } d > 0 \text{ and } r < 1$$
$$\iff \text{unstable polymorphism obtains at } p = \pi_A.$$

This establishes the correspondences in Table 5.12.

PART III

Rationality Meets Evolution

6

The Evolution–Rationality Connection

6.1 Introduction

This chapter and the next two examine the connection between evolution and rationality, in particular between evolutionarily optimal behaviour and rational behaviour. This connection is subtle and multi-faceted; aspects of it have been discussed by workers in a number of disciplines, including philosophy, behavioural ecology and economics.[1] My aim is not to survey all of this work, but rather to discuss key conceptual issues and to offer an overarching philosophical perspective.

The evolution/rationality connection has two distinct though related dimensions. First, there is the issue of the evolution of rationality itself. Given that creatures able to think and act rationally, including *Homo sapiens*, have evolved, what explains this? Should we think of rationality, or the capacity for it, as itself a biological adaptation, and if so, what selective advantage did it confer? Could evolution ever lead to irrationality? Second, there is the role of rationality concepts in evolutionary biology. We saw one instance of this in chapter 1: agential thinking (type 1), in which an evolved organism is treated as if it were a rational agent pursuing a goal, towards which its phenotypic traits, in particular its behaviour, conduce. Another is the use of models and concepts borrowed from rational choice theory, for example, game and decision theory, in evolutionary analysis. Thus rationality-inspired ideas feature in biology, even when the organisms under study have limited cognitive abilities.

One interesting hypothesis, which potentially connects these two dimensions, is that adaptive behavioural plasticity in non-human animals constitutes a sort of proto-rationality, in the sense of being functionally similar to the behaviour of an instrumentally rational agent (and probably an evolutionary precursor of it). An animal that exhibits such plasticity performs actions that are appropriate to its goal of surviving and reproducing, or maximize its fitness, given the information it has

[1] In philosophy, see Sober (1998), Skyrms (1995), Danielson (1998), and Cooper (2003); in behavioural ecology, Kacelnik (2006), and Houston *et al.* (2007a); in economics, Binmore (1987), Robson (2002), and Robson and Samuleson (2010).

about the environment, while an agent that behaves rationally chooses actions that are appropriate to their goal, or maximize their utility, given their beliefs. Indicative of this is that the fitness-maximizing paradigm in behavioural ecology and the utility-maximizing paradigm in rational choice theory are similar, formally and conceptually. Moreover, there is some evidence that belief-desire cognitive architecture, which is widely thought to underpin human rational action, is also found in animals.[2]

The idea of behavioural plasticity as proto-rationality links the two dimensions of our problem, for it both suggests a route by which rationality might have evolved and helps explain the suitability of rationality concepts for describing evolutionary adaptations. If fully fledged human rationality evolved from the behavioural plasticity of non-human animals, then we should expect a series of intermediate stages where rational descriptors are partially but not fully applicable. Moreover, since behvioural plasticity within an organism's lifetime achieves something similar to what evolution achieves across generations—fit of organism to environment—we should expect that if rationality concepts can illuminate the former, they can also illuminate the latter. This link between the two dimensions is mirrored in the fact that certain formal models admit of a dual interpretation: either as explanations of how rationality evolved, or of why rational idioms are well-suited to characterizing adaptive behaviour; an example is developed in section 6.5.

The first dimension above—the evolution of rationality—may seem like a purely scientific matter, similar in kind to asking about the evolution of any organismic attribute, for example, bone density. However, things are not so simple. For one, the concept of rationality can be understood in various ways, so the question of how rationality evolved does not have a single meaning. Moreover, rationality, however understood, is a *normative* notion—to call something rational is not just to describe it but also to evaluate it. The same is not true of all organismic attributes. This raises the philosophical question of whether an evolutionary explanation of rationality can tell us anything about its normative status. For example, might the rational/irrational distinction be reducible to the adaptive/maladaptive distinction?

Studying the evolution–rationality connection is complicated by the fact that rationality lacks a canonical definition. Section 6.2 introduces A. Kacelnik's useful distinction between the biological, economic, and philosophical-psychological concepts of rationality, and considers how they relate (Kacelnik 2006). Section 6.3 asks whether rationality is a biological adaptation, and argues that the answer may depend on whether rational is contrasted with *a*rational or *i*rrational. Section 6.4 considers the relation between the two dimensions of the evolution–rationality connection distinguished above. Section 6.5 sketches a simple formal model, for illustrative purposes, which suggests a biological basis for the norms of Bayesian rationality.

[2] See Carruthers (2006) for discussion of this point.

Section 6.6 compares the fitness concept in biology with the utility concept in rational choice theory, in respect of their theoretical roles. Section 6.7 asks whether rationality norms can be naturalized by evolutionary biology. Section 6.8 concludes.

6.2 Concepts of Rationality

In economics and rational choice theory, rationality is understood behaviouristically, as consistency of preference or choice, and given a mathematically precise meaning (e.g. Kreps 1988, Mas-Colell *et al.* 1995). In the simplest case of choice between certain outcomes, for example, commodity bundles, a rational agent's preferences are required to satisfy certain conditions, notably transitivity. (In principle this can be checked by offering the agent a sequence of binary choices.) These conditions imply that the agent behaves as if they have a utility function on the outcomes, and always prefers the outcome with the highest utility. Thus an agent is rational if and only if their choice behaviour can be characterized as utility-maximizing. Note that the utility function in this case is purely ordinal.

Economic rationality becomes more interesting in less simple cases. In choice under uncertainty, the outcome an agent gets from a given action depends on the state of the world, which they don't know. An action can thus be equated with a function from worldly state to outcome, that is, a specification of which outcome will be received in each state. An ingenious argument, due originally to Ramsey (1931) and Savage (1954), shows that as long as the agent's choices between actions satisfy certain fairly intuitive consistency conditions, then they behave as if they assign subjective probabilities to the states of the world, have a (cardinal) utility function on the outcomes, and choose between actions in order to maximize expected utility. This concept lies at the heart of Bayesian decision theory.

Another interesting case concerns inter-temporal choice, in which an agent must choose between present and future rewards. Rationality, on one standard and fairly plausible view of the matter, then requires that an agent not commit themselves to choices that they know they will later regret. Thus if the agent prefers $\$x$ today to $\$x+1$ tomorrow, they should prefer $\$x$ in a year's time to $\$x + 1$ in a year and a day—since otherwise, in a year's time they will regret their previous choice. This 'stationarity' condition, along with transitivity and other conditions, implies that the agent behaves as if they discount the utility of future rewards at a constant rate; this then gives rise to 'exponential discounting', the classic normative model of inter-temporal choice (Samuelson 1937, Fishburn and Rubinstein 1982).

Note that utility-maximization in economics is not a psychological hypothesis. There is no assumption that the agent has a numerical utility function in their head, or computes expectations or discount rates. Rather, the agent's preferences are assumed to satisfy certain axioms or rationality constraints, from which it follows as a mathematical consequence that the agent behaves as if trying to maximize utility (or expected utility or discounted utility). So 'utility' is in effect defined as whatever the agent appears to care about. Rationality in this sense is a property of an agent's choice behaviour, not of the psychological mechanisms that produce it.

Many philosophers define 'rationality' in a psychologically richer way, to mean that an agent has good reasons for their actions and beliefs, and that these reasons have led the agent to act/believe as they do.[3] Traditionally philosophers distinguish theoretical rationality, which deals with principles of rational belief, from practical rationality, which deals with principles of rational action. Examples of the former include logical consistency, probabilistic coherence, and taking account of all relevant evidence; of the latter, choosing actions likely to further one's desires given one's beliefs, or what is often called 'instrumental rationality'. Understood in the traditional philosophical way, rationality applies only to organisms who can reason about what to believe and do; whether non-humans qualify as rational in this sense is controversial. By contrast, conforming to the consistency conditions of rational choice theory could be achieved by an organism who lacked reasons but was capable of making behavioural choices. Indeed researchers have studied whether the choice behaviour of rats, pigeons, and even insects satisfies the expected utility and exponential discounting models.[4]

In a useful discussion, Kacelnik (2006) refers to rationality in the sense of acting or believing on the basis of reasons as P-rationality (for philosophy and psychology), by contrast with the E-rationality of economists.[5] He also describes a third notion found in evolutionary biology, B-rationality, which is based on adaptive considerations. Essentially, an organism's behaviour is B-rational if it is adaptive or biologically optimal, that is, if it maximizes fitness, or some component thereof, in the relevant environment. Such behaviour is rational in that it furthers the organism's 'biological goal' of surviving and reproducing.

A taxonomy based on disciplinary divisions might be thought unlikely to carve nature at its joints; and Kacelnik's does gloss over certain complications. One is that there are ongoing debates surrounding each of the three concepts he identifies. In the case of E-rationality, not all theorists agree on what the 'right' axiomatic conditions are; for example, some regard the expected utility axioms as rationally compelling, others as overly demanding, or applicable only to well-defined decision problems. In the case of P-rationality, some philosophers take instrumental rationality to exhaust the meaning of rational action, while others think further constraints are necessary, for example, that a subject's goals are ones that on reflection they would wish to have. In the case of B-rationality, the fitness concept does not have a univocal definition in biology, as discussed in section 1.7. So matters are less simple that Kacelnik implies; but nonetheless his tripartite distinction is a real one.

A final complication concerns the relation between P- and E-rationality. Many philosophers regard expected utility theory, which belongs to E-rationality, as a

[3] Recall the discussion of different concepts of agency from section 1.2.
[4] See Kagel *et al.* (1995) and Kalenscher and van Wingerden (2011).
[5] Kacelnik's actual expression is PP-rationality, but I simplify this to P-rationality.

formalization of everyday belief-desire psychology, and thus as relevant to P-rationality (e.g. Lewis 1981). On this view, maximization of expected utility is a formal expression of the core notion of instrumental rationality, that is, choosing an action that one believes will bring an outcome one wants. Proponents of this view typically oppose the behaviouristic reading of the expected utility model, and attempt to imbue the utility and probability functions, which fall out of Savage's representation theorem, with psychological meaning. I have argued elsewhere that this is unhelpful, as it misconstrues the normative message of expected utility theory, which concerns behaviour, not psychological process (Okasha 2016). However, the ongoing debate on this point does muddy the distinction between E- and P-rationality.[6]

Note that B-rationality is an externalist concept, since the fitness consequences of a given behaviour or strategy depend on the environment. By contrast, rationality in the sense of having reasons for one's beliefs and actions, or in the sense of making choices in accordance with consistency conditions, are internalist matters. Whether an agent is rational in these senses depends on how things are in their head, not in the external environment; so a suitably intelligent agent should be able to achieve rationality by self-reflection and amelioration. For biological rationality, the world must cooperate too.

How do Kacelnik's three rationality concepts relate? A natural thought is that they describe different aspects of a single phenomenon. E-rationality describes the conditions that behaviour must satisfy in order to appear goal-directed, or to maximize some quantity or other. P-rationality describes the psychological processes that produce such behaviour in cognitively sophisticated organisms, who can mentally represent their goals. B-rationality describes the special case of E-rationality when the quantity that the behaviour maximizes is Darwinian fitness. I think this is partly right, but not the whole story.

McFarland (2016) queries whether B-rationality deserves its name, on the grounds that it is simply biological adaptiveness in another guise. There is some merit to this objection. After all, for any trait with a biological function, for example, the mammalian heart, we can ask whether it is functioning as it should; but this is not a species of rational evaluation, and to treat it as such may seem unhelpful. However, in Kacelnik's defence, matters are different where evolved behaviour is the trait of concern, especially if an organism can learn about the environment and modify its behaviour accordingly. Such behaviour is functionally similar to that of rational human behaviour—flexible, goal-directed, environmentally sensitive—and is presumably its evolutionary precursor. Indicative of this is that the scientific study of such behaviour does draw on rationality concepts, as we saw in chapter 1.

To illustrate, consider optimal foraging theory in behavioural ecology (Stephens and Krebs 1987). A classic problem in this field is foraging in a patchy environment,

[6] See Buchak (2013) and Dietrich and List (2016) for discussion of this point.

in which an organism must decide when to move from one patch to another as its food intake declines. A typical study might ask whether a bird's decision-rule for moving patch maximizes its probability of getting enough food to survive the night. Essentially, this is to ask whether the bird's evolved decision-rule, combined with the information it acquires from the environment, enables it to make behavioural choices that are appropriate to its goal of maximizing survival. The word 'rational' is not always used in such studies, but a species of rational evaluation is arguably at work; and foraging models often treat the organism as a Bayesian agent, updating its beliefs about patch quality by conditionalizing on new information.[7]

Despite this consideration, I avoid the expression 'B-rational' below, but for a different reason from McFarland's. For it invites the implication that adaptive behaviour is simply equivalent to E-rational behaviour, with utility equated to biological fitness. This is indeed how Kacelnik sees the matter; however, I think of this as an interesting hypothesis to be explored, not a definitional truth; and as we shall see in chapter 7, many putative counterexamples to the 'utility = fitness' hypothesis have been suggested. Thus I employ Kacelnik's terminology of E- and P-rationality, but use 'adaptive' or 'biologically optimal' in lieu of his B-rationality.

6.3 Rationality as Evolutionary Adaptation

The label 'rational', in any of its senses, has two contraries: irrational and arational. The former means that the relevant rationality requirements could have been met but weren't; the latter that they were not applicable in the first place. We can thus distinguish two questions about the evolution of rationality. The first concerns the transition from arationality to rationality. How did organisms evolve the behavioural and cognitive capacities necessary for the notions of E- and P-rationality to be applicable to them at all? The second concerns the adaptive significance of being rational as opposed to irrational. Once the relevant capacities were in place, would natural selection have favoured organisms that satisfy the respective rationality criteria?

6.3.1 From arationality to rationality

What are the prerequisites for E- and P-rationality to apply? E-rationality requires that an organism can choose between alternatives. Though in a sense any behaviour can be described as involving choice, for example, a plant 'chooses' where to grow its roots, I think that E-rationality is only usefully applied to organisms with nervous systems that exhibit activational plasticity (Snell-Rood 2013). This means that the organism can modify its behaviour adaptively, and reversibly, in response to external stimuli in real time; the contrast is with developmental plasticity, in which an organism's developmental trajectory is influenced by the environment. One could hardly perform

[7] See for example Green (1980).

a forced-choice experiment on a microbe or plant to test for violations of transitivity, for example, as one can for a bird or mammal.

P-rationality requires that an organism have beliefs and desires, and can reason, for example, perform practical inferences of the form 'in order to get *x*, I should do *y*'. Some would restrict rationality of this sort to humans on the grounds that language is required to have beliefs, but this is questionable.[8] A more plausible view is that any organism that is capable of learning may be said to have beliefs, or at least proto-beliefs (Dretske 1998). This weaker criterion for belief, combined with the fact that many animal researchers hold that their subjects' behaviour requires explanation in intentional terms (e.g. Clayton *et al.* 2006), and the evidence of continuity between humans and animals in respect of instrumental reasoning (MacIntyre 1999, Carruthers 2006), suggests that some non-human organisms exhibit the cognitive capacities needed for P-rationality to be applied meaningfully. I think of this not as an empirical hypothesis, but rather as an empirically informed decision about how to use certain concepts.

To understand the transition from arationality to rationality, therefore, we need to know why evolution led to the capacities—activational plasticity, learning, belief- and desire-like representations—that are the preconditions for P- and E-rationality to apply. In very broad terms, the answer seems clear. These capacities enhance an organism's ability to exhibit phenotypic plasticity, that is, varying its phenotype depending on the context in which it finds itself. A long tradition in evolutionary biology teaches that phenotypic plasticity is adaptive in a variable environment.[9] If the environment is fixed, natural selection can simply hard-wire in the appropriate phenotype, but if it varies, either temporally or spatially, an organism will do better if it can adjust its phenotype suitably, in response to an environmental cue. This general point applies widely, and covers plasticity of all sorts, morphological and physiological; behavioural plasticity and learning are special cases of it.

Though the adaptive benefit of behavioural plasticity is fairly clear, this alone does not explain why some organisms evolved the ability to represent mentally the environment and their own goals, and to use those internal representations to guide their behaviour. For the fact is that many organisms exhibit adaptive behaviour, which is plastic to a greater or lesser extent, without these cognitive attributes. This point has been emphasized by Barrett (2011), who argues that simple internal mechanisms can often produce surprisingly complex behaviour by taking advantage of environmental regularities. Barrett gives the example of predatory *Portia* spiders, which show a remarkable ability to detour around obstacles while hunting for prey. The spiders' behaviour conveys the impression of advance planning, for they need to move out of sight of their goal to get around an obstacle, and appear to scan the terrain carefully before setting off on a detour. However, ingenious experiments reveal that no planning

[8] The view that beliefs require language is defended by Davidson (2001a, b). For effective rebuttals, see McIntyre (1999), Kornblith (2002), and Rescorla (2013).

[9] See Whitman and Agrawal (2009) for a review.

is going on; the spider is using a simple rule of thumb, based on the presence or absence of horizontal lines in its field of vision, to determine which direction to move in at each moment; this leads it to trace out the most direct route towards its prey. Such examples, of which there are many, show that flexible behaviour can evolve without the cognitive capacities that are requisites for the concept of P-rationality to apply.

Barrett's point is well-taken; however, I regard it as compatible with the hypothesis that the function of cognition in general, and of mental representations in particular, is to facilitate adaptive behaviour in novel environments. For the organisms in Barrett's examples display adaptive behaviour only within fairly narrow environmental bounds, as she rightly stresses. The *Portia* spider's flexibility is limited to particular activities, such as hunting, and works well in the forests where the spider hunts. It remains plausible that behaving adaptively across a range of novel environments requires having internal representations, that is, belief- and desire-like states, which guide behaviour. A detailed defence of this view is offered by Godfrey-Smith (1996), who argues that the function of cognition is to allow an organism to cope successfully with environmental complexity.

One version of this view holds that the complexity of the *social* environment in particular, for example, interactions with con-specifics, is what drove the evolution of hominid cognition (Byrne and Whiten 1988). A persuasive defence of this view is offered by Sterelny (2003), who argues that 'translucent environments', in which cues are noisy, select for what he calls 'decoupled representation': internal belief-like states that can be used by the organism in the service of many different actions, rather than being rigidly tied to a single action. Thus belief- and desire-like representations evolve because they enable an organism to display adaptive behaviour in the face of social complexity, Sterelny argues.

These brief remarks are not intended to settle the debates surrounding the evolution of behavioural plasticity, learning, or belief- and desire-like representations. There is a real question about why these attributes evolved in some lineages but not others, and what the precise sequence of intermediate stages was like. However, it seems very plausible that these attributes did evolve by natural selection, and in broad terms their biological significance is clear: facilitating adaptive behaviour where natural selection cannot hard-wire it. Thus the empirical question of why rationality evolved from arationality, that is, why organisms evolved the capacities necessary for evaluation by the criteria of E- and P-rationality to be meaningful, can be satisfactorily answered, I suggest.

6.3.2 *Selection for being rational?*

Our second question concerns the evolution of rationality as contrasted with irrationality, rather than arationality. Would organisms that satisfy the standard norms of E- and P-rationality, presuming that the preconditions for these norms to apply are met, have enjoyed an evolutionary advantage? The norms in question include consistency of choice, coherence of beliefs, valid reasoning, and performing actions well-suited to achieving one's goal.

A preliminary question is whether organisms *do* in fact satisfy these norms. This cannot be taken for granted, for there is evidence that both humans and animals systematically violate certain rationality norms, at least in some contexts. In humans, behavioural economists and psychologists have uncovered numerous errors of reasoning and anomalies of choice; these include preference reversals, time-inconsistency, violations of the expected utility axioms, uncertainty aversion, loss aversion, probabilistic incoherence, the base-rate fallacy, framing effects, and more (Camerer 2003, Kahneman 2011). In animals, experimenters have found intransitive choices, expected utility violations, time-inconsistency, and more (McFarland 2016, Kagel *et al.* 1995, Houston 2012). Interestingly, the violations of E-rationality in animals and humans are qualitatively quite similar.

The correct interpretation of these experimental data is moot. One interpretation takes the data at face value: humans and animals are simply less than fully rational, for whatever reason. An alternative interpretation is that the experimental scenarios are ecologically unrealistic, that is, the anomalies arise because organisms' behavioural/psychological dispositions are being studied in an artificial setting.[10] A third, complementary interpretation is that the wrong normative standard is being applied. This is argued for in the human case by G. Gigerenzer and colleagues in their work on 'ecological rationality' (Gigerenzer and Selten 2001, Gigerenzer and Todd 1999, Gigerenzer *et al.* 2011). Gigerenzer claims that the norms of probability and decision theory are an inappropriate benchmark against which to judge evolved humans; we rely on 'adaptive heuristics' to reason and make decisions, and these typically work well in the relevant environments, he argues.

Rather than trying to adjudicate this issue, which is partly an empirical one, let us take a step back and ask whether we should expect natural selection to favour satisfaction of E- and P-rationality norms in the first place. Would organisms who choose actions well-suited to their goals, whose reasoning obeys the laws of logic and probability, and whose choices are internally consistent have enjoyed a selective advantage? Such claims are often made in the philosophical literature. For example, Dennett (1987) writes: 'natural selection guarantees that most of an organism's beliefs will be true, most of its strategies rational' (p. 7); W. V. Quine (1969) argues similarly, as does R. de Sousa (2007). However, a detailed defence of this claim has rarely been offered; and the systematic departures from ideal rationality found in both humans and non-humans dictate caution on the point.

Clearly, being E- or P-rational will only tend to enhance an organism's biological fitness if it has preferences or desires that are biologically appropriate, that is,

[10] This point has recently been argued by Fawcett *et al.* (2014).

suitably fitness-enhancing. Having consistent preferences for things that are biologically irrelevant or harmful, or reasoning correctly about how to get them, will do little for fitness. However, modern humans often have such preferences and desires, using their intelligence and rationality to pursue goals that are unrelated to, and sometimes inimical for, biological fitness. As Sterelny (2012) notes, at some point in hominid evolution, we ceased to 'want what our genes want' (p. 263).

This familiar point does not refute the idea of rationality as an evolutionary adaptation, for two reasons. First, having preferences and desires that are divorced from biological fitness is a phenomenon unique to humans, hence phylogenetically recent; it may well be an unselected side-effect of our increased cognitive capacity, or the result of cultural evolution, or both. Second, it is plausible that however the phenomenon arose, it altered our motivations but left intact our basic cognitive architecture, including our ability to think, reason, and choose. Evidence for this is that we deploy our capacity for instrumental rationality in essentially the same way whether we are pursuing a biologically relevant goal, for example, finding a mate, or a harmful one, such as climbing a sheer rock face for fun.

Although E-rational behaviour will not necessarily be adaptive, the converse inference has often been defended (e.g. Gintis 2009, p. 7, Kacelnik 2006, Chater 2012). The inference seems sound: if an organism always makes the fitness-maximizing choice, then presumably it is behaving like a utility-maximizing agent whose utility function is its biological fitness? Thus it seems that if an organism's behaviour is adaptive, it will automatically be utility-maximizing or E-rational. I think this will often be true. However, a number of authors have argued that in certain contexts, natural selection can favour behaviours that *violate* the consistency criteria of E-rationality, for example, intransitive choices. These arguments, surprising at first sight, are scrutinized in chapter 7.

Behaviour that is P-rational need not be adaptive either, as the example of humans with biologically aberrant desires shows. However, it is presumably true that *if* an organism has biologically appropriate desires, and *if* it has true beliefs, then being instrumentally rational, that is, acting in ways that it believes will lead to the satisfaction of its desires, will be selectively advantageous. Conversely, if an organism is instrumentally rational and has appropriate desires, then having true beliefs will be advantageous; and if an organism is instrumentally rational and has true beliefs, having appropriate desires will be advantageous.

These obvious points suggest that the instrumental rationality component of P-rationality, at least, may have a Darwinian explanation. If the outcome of a given action depends on the environment, as is usually the case, then one way natural selection can induce an organism to perform a biologically optimal action is to equip it with true (and therefore consistent) beliefs about the world, or the ability to form such beliefs by learning, with desires for outcomes that correlate suitably with biological fitness, and with instrumental rationality. A Bayesian version of this argument is elaborated in section 6.5.

Of course, this argument does not show that the *only* way for natural selection to induce biologically optimal behaviour is by making the organism instrumentally rational and equipping it with true beliefs and biologically appropriate desires; so at most it shows how instrumental rationality and consistency of belief *could* have evolved. It is also conceivable that optimal behaviour, at least in typical environments, could be produced by cognitive mechanisms that are psychologically irrational. For on general grounds, we should expect that a single pattern of behaviour could be produced by a variety of cognitive processes.

This last point is again illustrated by Gigerenzer's work. He argues that much human cognition is irrational as judged by traditional norms, for example, involves faulty inferences, short-cuts, flawed probability assessments, and a failure to assess evidence correctly, but nonetheless works well in the natural environments in which it are used. By 'works well' Gigerenzer does not necessarily mean achieve high fitness—his focus is mostly on cognitive rather than biological criteria of success—but the point extrapolates to the latter. If agents who violate traditional rationality norms nonetheless perform the 'right' actions, then natural selection will not discriminate against them. The point is that natural selection acts on behaviour, not directly on the cognitive processes that give rise to behaviour.

Is rationality then an adaptation? I think there is almost certainly a Darwinian explanation for why some organisms are rational rather than arrational, that is, why they evolved the capacities that are prerequisites for rationality criteria to apply. Whether there is also a Darwinian explanation for why these organisms are rational rather than irrational, to the extent that they are, is less clear. Certainly, a plausible case can be mounted for why satisfying traditional rationality criteria (of both the E and P varieties) will often tend to enhance biological fitness, but this case is only suggestive. Indeed it is conceivable that there is a Darwinian explanation for why organisms, including humans, sometimes *violate* traditional rationality criteria.

6.4 Interlude: Relating the Two Dimensions

Recall the distinction between the evolution of rationality and the use of rationality concepts, as heuristics or metaphors, in evolutionary analysis. In the light of our discussion so far, it is worth considering how these relate.

Suppose it is true that the prerequisites for E- and P-rationality—behavioural plasticity, learning, belief- and desire-like representations—evolved in order to permit real-time adaptive responses to the environment, as we have suggested. In effect, this means that the transition from arationality to rationality involved a transfer of choice-making power from natural selection to the individual organism. Prior to the evolution of these capacities, natural selection chose between alternative phenotypes according to how well they performed in a fixed environment. As environments complexified, organisms evolved the ability to learn, make choices, and engage in flexible goal-directed activity. Thus natural selection's choices, made over inter-generational time, eventually gave rise to organisms capable of real-time choice—and in the hominid lineage, to pursue goals that are divorced from biological fitness and engage in sophisticated reasoning about how to achieve them.

This is a theory (or story) about how the capacities necessary for rationality evolved; but it also has implications for the use of rationality-inspired heuristics in evolutionary biology. In particular, it helps explain why agential thinking, the intentional idiom, and the organism-as-agent concept are useful for understanding adaptation, even in organisms that lack cognitive sophistication. The point is not simply that the capacities in question are themselves adaptations; the same is true of many organismic attributes. Rather, the point is that the evolved function of the capacities—what they have been designed for—is to enable an organism to exhibit adaptive behaviour in its lifetime in response to environmental contingencies, and this is functionally equivalent to what the process of Darwinian evolution itself achieves when it adapts an organism to the environment over inter-generational time. Thus it becomes intelligible why a set of concepts and descriptors whose original application is to the purposive behaviour of humans should be heuristically useful in adaptationist theorizing.

This helps answer a puzzle posed in this book's Introduction. We observed that the biological practice of treating evolved organisms as if they were agents with interests and goals was in some ways quite odd. After all, it is not heuristically useful to treat invertebrates as if they had backbones, so why should it be useful to treat organisms that lack the attributes of rational agency as if they had them? If the story sketched here is correct, the resolution of the puzzle is as follows. Certain behavioural and cognitive capacities distinguish rational from arational creatures; and the biological function of those capacities is to permit adaptive responses in complex environments. Such within-lifetime responses are one way of becoming well-adapted to the environment; another is Darwinian evolution over many generations. These modes of adaptation have much in common; hence it can be useful to treat organisms that lack the capacities in question as if they had them. That is the fundamental reason, I suggest, why the organism-as-rational-agent heuristic is useful in evolutionary biology, rather than being mere anthropomorphism.

On the other hand, a fuller vindication of the heuristic would require that traditional rationality norms, rather than just the capacities that those norms presuppose, have an evolutionary explanation; and we have suggested that this is a taller order. For suppose it turns out that there is an evolutionary explanation for why organisms sometimes violate those norms, for example, probabilistic coherence or transitivity of choice, as some authors suggest. This would imply that the criteria of biological adaptiveness and rationality can fail to coincide, which would undermine the use of rationality concepts to describe or theorize about the phenomenon of Darwinian adaptation. We cannot usefully treat an evolved organism as akin to a rational agent trying to achieve a goal, if the organism's evolved behaviour violates the norms of rationality that this metaphor invokes. Therefore, arguments to the effect that natural selection may sometimes favour irrationality, if correct, would also show that the organism-as-rational-agent heuristic has its limits. Such arguments are explored in chapter 7.

6.5 Evolution of Bayesian Rationality?

This section develops a simple model of the evolution of Bayesian rationality, by considering optimal behaviour in the face of uncertainty. The model is not biologically realistic but serves to fix ideas. The aim is to show how Bayesian norms emerge naturally in a biological setting, and to provide a touchstone for the later discussion of how irrationality can evolve. The model also highlights the interplay between the notions of E-rationality, P-rationality, and biological adaptiveness, in a simple case.

Bayesian rationality involves three components: (i) partial beliefs or credences that conform to the probability calculus; (ii) updating of beliefs by conditionalization; and (iii) choice between options in accordance with expected utility maximization. One standard use of (i)–(ii) is to model the beliefs and choices of rational human agents; however, they have also been applied in behavioural ecology.[11] To readers familiar with Bayesian epistemology this may seem surprising, as non-human organisms might be thought incapable of having probabilistic beliefs, less still of updating them; so it merits brief explanation.

A typical Bayesian approach in behavioural ecology is to assume that organisms have some information about an environmental parameter, represented by a probability distribution. For example, an organism may 'know' that food patches are of two types, good and bad, with equal frequency. Talk of knowledge here is intended behaviouristically: it means that the organism behaves as if it had the information in question. Such information may either be hard-wired into the organism's genome, acquired by experience, or both. There is abundant evidence that organisms of all taxa make adaptive use of information about their environment (Giraldeau 1997); and some evidence that organisms process information in a Bayes-like manner, that is, they combine their prior knowledge with information acquired from experience to produce an updated knowledge state, which then informs their behaviour (Valone 2006).

This behaviouristic understanding of Bayesian principles is out of favour in modern epistemology, but is consonant with the approach of decision theorists in the tradition of Savage (1954), who derives an agent's subjective probabilities (and utilities) from their choices between uncertain options, which are in principle observable. This approach is orthodox in economic theory, if not in epistemology. So the biologists' use of Bayesian ideas to describe adaptive behaviour is not such a radical departure. The issue of whether (i)–(iii) are to be interpreted behaviouristically, as part of

[11] See for example McNamara *et al.* (2006), Valone (2006).

E-rationality, or mentalistically, as an idealized model of P-rationality, is revisited in section 6.3. Our first task is to show how behaviour that conforms to Bayesian principles arises naturally in a simple biological setting.

6.5.1 *Motivating example*

An organism is foraging for food in a predator-strewn area. Predators are of two types: snakes and leopards. If no predator is present, the best thing to do is stay put and forage. If a snake is present, the best thing to do is to climb a tree. If a leopard is present, the best thing to do is flee. Both climbing and fleeing are costly in terms of time and energy, to different extents. Payoffs for the three actions, as a function of the state of the world, are shown in Table 6.1; these payoffs are measured in increments of biological fitness, that is, number of offspring.

What action should the organism choose to maximize its expected payoff? That depends on the probabilities of the three states of the world. Suppose that the probabilities of the three states are:

$$\Pr(\text{no predator}) = 1/2, \Pr(\text{snake}) = 1/3, \Pr(\text{leopard}) = 1/6$$

(These probabilities can be thought of as the relative frequencies with which each state occurs in the relevant environment.) Then, the expected payoff from each action is:

$$\begin{aligned} V(\text{stay put}) &= 10(1/2) + 0(1/3) + 0(1/6) = 5 \\ V(\text{climb}) &= 5(1/2) + 4(1/3) + 0(1/6) \approx 3.83 \\ V(\text{flee}) &= 6(1/2) + 1(1/3) + 2(1/6) \approx 3.67 \end{aligned}$$

So the optimal action, as judged by the criterion of expected reproductive success, is to stay put. Organisms choosing to stay put will on average leave more offspring than those choosing either of the other two actions. If the organism's choice behaviour has been optimized by natural selection, or if it has learnt which action is optimal, then it will stay put.

Now suppose that prior to choosing an action, the organism receives a signal which indicates whether a predator is present or not. So the signal has two values: 'safe' and 'unsafe'. The signal is perfectly reliable, indicating 'unsafe' if and only if a predator is present, but cannot discriminate between leopards and snakes. The organism's choice of action may depend on which signal is received. So the organism needs to have a

Table 6.1. Payoffs for three alternative actions

	No predator	Snake	Leopard
Stay put	10	0	0
Climb	5	4	0
Flee	6	1	2

'policy', that is, a specification of which action to take for each value of the signal. There are 9 (=3^2) possible policies; we consider three in particular.

Suppose first that the organism ignores the signal, and always chooses the action optimal for the situation in which no signal is received. So the organism's policy is 'if safe, stay put; if unsafe, stay put'. Call this policy *Ignore*. Such a policy may seem unpromising, but it could reflect perceptual or cognitive limitations. If an organism is incapable of attending to the signal, or lacks behavioural plasticity, it might use a policy like *Ignore*.

Second, suppose that the organism does attend to the signal and adopts a maximin strategy, choosing the action that maximizes the minimum payoff it will receive, in the light of the information provided by the signal. So if the safe signal is received it chooses to stay put. If the unsafe signal is received it chooses to flee, as this guarantees it a payoff of at least 1. So its policy is 'if safe, stay put; if unsafe, flee'. Let us call this policy *Maximin*. The policy reflects a high degree of risk-aversion. By choosing to flee rather than climb when the unsafe signal is received, the organism forgoes a possible payoff of 4 in order to definitely avoid a payoff of 0.

Third, suppose that the organism behaves like a Bayesian agent. That is, the organism has probabilistic beliefs ('credences') about the state of the world which it updates by conditionalization, a utility function over the outcomes, and chooses between action so as to maximize expected utility relative to its current beliefs. Suppose that the organism's initial credences match the true frequency distribution on the states, and that its utility function equals its fitness payoff. So if the safe signal is received, the organism stays put—as the conditional expected payoffs are then 10, 5, and 6 for stay put, climb, and flee, respectively. What if the unsafe signal is received? The conditional probabilities of the three states, given an unsafe signal, are 0, 2/3, and 1/3, respectively. The conditional expected payoffs for staying put, climbing, and fleeing are then 0, 8/3, and 4/3, respectively, so the organism climbs. Its policy is therefore 'if safe, stay put; if unsafe, climb'. Let us call this policy *Bayes*.

Which of our three policies will natural selection favour? To address this, we compute the expected payoffs accruing to an organism that uses each. Consider first *Ignore*. Since an organism using *Ignore* chooses to stay put whether or not a signal is received, its expected payoff is:

$$V(\mathit{Ignore}) = 10(1/2) + 0(1/3) + 0(1/6) = 5$$

What about *Maximin*? An organism using *Maximin* stays put if a safe signal is received but flees otherwise. With probability 1/2 no predator is present, so the safe signal is sent, so the organism stays put and earns a payoff of 10. With probability 1/3 a snake is present, so the unsafe signal is sent, so the organism flees and earns a payoff of 1. With probability 1/6 a leopard is present, so the unsafe signal is sent, so the organism flees and earns a payoff of 2. Its expected payoff is therefore:

$$V(\mathit{Maximin}) = 10(1/2) + 1(1/3) + 2(1/6) \approx 5.67$$

By a similar logic, the payoff to an organism using *Bayes* is:

$$V(Bayes) = 10(1/2) + 4(1/3) + 0(1/6) \approx 6.33$$

Therefore, *Bayes* yields the highest expected payoff. If all three policies are found in a population, natural selection will favour *Bayes* over the other two. This suggests, in a preliminary way, that evolution should produce organisms that behave as if they were Bayes-rational agents, with subjective credences that match the objective frequencies (so are 'perfectly calibrated'), who respond rationally to new information, and who make rational choices. A stronger argument for this conclusion would need to show that no conceivable policy does better than the *Bayes* policy; this is shown below.

6.5.2 *Can't do better than Bayes*

The framework we adopt directly generalizes the previous example. There is a finite set S of states of nature; $S = \{\theta_1, \ldots, \theta_k\}$; the probability of state θ_i, understood as its frequency of occurrence, is $p(\theta_i)$. There is a finite set A of actions that the organism may perform. The payoff from an action depends on the state of nature. The payoff from action $a \in A$ in state of nature θ_i is $V_i(a)$. As before, payoffs are measured in increments of biological fitness.

There is a finite set E of possible signals that the organism may receive; $E = \{E_1, \ldots, E_n\}$. In the previous example we assumed that the signal set partitioned the states of nature, that is, each state was compatible with exactly one signal; here we relax that assumption. If the state of nature is θ_i, the probability of receiving signal E_j is $p(E_j|\theta_i)$. Thus the marginal probability of signal E_j is $p(E_j) = \sum_S p(\theta_i) \cdot p(E_j|\theta_i)$. We assume that $p(E_j) > 0$ for all j, that is, every signal occurs with non-zero probability.

An organism's policy specifies an action $a \in A$ for every possible signal $E_j \in E$. Thus a policy is simply a function from E to A. The set of all policies is denoted X. For any policy $x \in X$, we let $x(E_j) \in A$ be the action specified by that policy when signal E_j is received; for convenience, we will write $x(E_j)$ as x_j.

Consider an arbitrary policy $x \in X$. What is an organism's expected payoff from using policy x? Suppose first that the true state of nature is θ_i. Then, the expected payoff from policy x equals:

$$\sum_{E_j} p(E_j|\theta_i) \cdot V_i(x_j) \qquad (6.1)$$

The justification for expression (6.1) is clear. $p(E_j|\theta_i)$ is the probability that signal E_j is received given that the state of nature is θ_i. $V_i(x_j)$ is the payoff from choosing action x_j—the action specified by policy x when the signal received is E_j—in state of nature θ_i. Expression (6.1) is thus the expected payoff to an organism using policy x when the true state is θ_i. Taking the expectation of expression (6.1) across states of nature thus gives us the overall **expected payoff to policy** x:

$$\sum_{\theta_i} p(\theta_i) \sum_{E_j} p(E_j|\theta_i) \cdot V_i(x_j) \qquad (6.2)$$

Now consider an organism that implements its policy in the following generic way. The organism has a prior credence over the states $c(\theta_i)$ and a utility function U on the action-state pairs; $U_i(a)$ is the utility from action a in state θ_i. On receipt of the signal E_j, the organism changes its credence by applying an update rule R to yield a new credence $R(c, E_j)$, then applies a decision rule D which specifies which action(s) in A to perform in the light of its updated credence. Thus $D(R(c, E_j)) \subset A$ denotes the set of action(s) that may be chosen by an organism with prior c who receives signal E_j and uses update rule R. If this set is a singleton for all E_j, then the quadruple $< c, U, R, D >$ picks out a unique policy, and otherwise a class of policies.

An organism of this generic sort is *Bayes-like* if the following hold: (1) its prior credence matches the true distribution on the states; (2) its utility function is its biological fitness; (3) its update rule is Bayesian conditionalization; and (4) its decision rule is expected utility maximization. See Table 6.2 for the formal definitions. The policy of a *Bayes-like* organism is called a *Bayes* policy; note that it need not be unique.

Note that the four conditions in Table 6.2 together imply:

$$D(R(c, E_j)) = \max_{a \in A} \sum_{\theta_i} p(\theta_i | E_j) \cdot V_i(a) \tag{6.3}$$

That is, the chosen actions of an organism with a *Bayes* policy, given signal E_j, maximize the conditional expected fitness across states of nature.

We wish to show that any *Bayes* policy is biologically optimal, that is, earns expected fitness no less than any other policy. The argument is very simple.[12] Let b be a *Bayes* policy; thus b_j is the action specified by b when signal E_j is received. By definition, we have $b_j \in D(R(c, E_j))$. Thus from equation (6.3), we have:

For every signal E_j,

$$\sum_{\theta_i} p(\theta_i | E_j) \cdot V_i(b_j) \geq \sum_{\theta_i} p(\theta_i | E_j) \cdot V_i(a) \quad \text{for all actions } a \in A \tag{6.4}$$

Table 6.2. A *Bayes-like* organism

$c(\theta_i) = p(\theta_i)$	'prior credences match frequencies'
$U_i(a) = V_i(a)$	'utility equals fitness'
$R(c, E_j)(\theta_i) = c(\theta_i \| E_j)$	'update rule is conditionalization'
$D(R(c, E_j)) = \max_{a \in A} \sum_{\theta_i} R(c, E_j)(\theta_i) \cdot U_i(a)$	'decision rule is EU maximization'

[12] The argument is a close relative of an argument, due originally to Brown (1976), that says that to maximize expected utility, an agent must change their beliefs by conditionalization.

Now consider an arbitrary policy x. Applying inequality (6.4) to the particular action $x_j \in A$ gives:

$$\sum_{\theta_i} p(\theta_i|E_j) \cdot V_i(b_j) \geq \sum_{\theta_i} p(\theta_i|E_j) \cdot V_i(x_j) \tag{6.5}$$

Applying Bayes' theorem and dividing across by $p(E_j)$:

$$\sum_{\theta_i} p(\theta_i) \cdot p(E_j|\theta_i) \cdot V_i(b_j) \geq \sum_{\theta_i} p(\theta_i) \cdot p(E_j|\theta_i) \cdot V_i(x_j) \tag{6.6}$$

Summing over all the signals, then reversing order of summation:

$$\sum_{\theta_i} p(\theta_i) \sum_{E_j} p(E_j|\theta_i) \cdot V_i(b_j) \geq \sum_{\theta_i} p(\theta_i) \sum_{E_j} p(E_j|\theta_i) \cdot V_i(x_j) \tag{6.7}$$

But the left-hand side of inequality (6.7) is the expected payoff to policy b, while the right-hand side is the expected payoff to policy x (expression (6.2)). So policy b is biologically optimal. Since nothing has been assumed about b except that it is a *Bayes* policy, we can conclude that every *Bayes* policy is optimal.

6.5.3 *Discussion*

What exactly does the above argument show? Using a standard criterion of evolutionary success—expected reproductive output—the model shows that evolution will favour an organism who behaves like a Bayes-rational agent. The organism will behave as if it has a subjective prior on the states which matches the true distribution; a utility function on the outcomes which equals biological fitness (or some positive linear transformation thereof); an updating rule which is Bayesian conditionalization; and a decision rule which is expected utility maximization. Taken together, these guarantee that the organism will implement a biologically optimal policy. Thus the key elements of Bayesian rationality appear to fall out of a simple biological model. (My reason for saying 'appear' will become clear in chapter 7.)

Note that the model assumes nothing about the organism's cognitive capacities. It assumes that the organism exhibits behavioural plasticity, and is capable of modifying its behaviour in response to a signal, but not that it has representational mental states, or belief-desire cognitive architecture, or the ability to engage in instrumental reasoning. So the model demonstrates an evolutionary basis for Bayesian rationality if the latter is construed in a strictly 'as if' way, in the manner of E-rationality and the traditional behaviouristic construal of decision theory. However, the model says nothing directly about P-rationality, since the prerequisites for this rationality concept to apply are not necessarily in place.

Of course, the organism *might* implement its policy, that is, function from states to actions, by having internal belief-like and desire-like states which combine to determine its choice of action, and by updating its belief state in response to new information. There is some evidence of such cognitive architecture in humans and some non-humans alike, as discussed previously. If so, then the model can be taken as showing that satisfying Bayesian norms, where these are construed as rationality requirements on mental states rather than on choice behaviour, is sufficient for arriving at a biologically optimal policy, as long as the agent's prior probabilities match the frequencies, and its utility function depends suitably on payoff, that is, biological fitness.

Taken this way, the argument is essentially a probabilistic version of the point made in section 6.3.2, namely that *if* an organism has true beliefs, desires that correlate suitably with fitness, and is instrumentally rational, then its resulting behaviour will be fitness-maximizing, or biologically optimal. However, the argument goes beyond this simple point, by showing that in addition the organism needs to adjust its probabilistic beliefs in a particular way, that is, by conditionalization, in order to achieve optimal behaviour. The caveat noted in section 6.3.2—that this is just one way of achieving fitness-maximizing behaviour—applies equally here. For an organism might implement the optimal policy by utilizing a simple heuristic or rule of thumb, even if it lacks representational states and belief-desire cognitive architecture altogether.

The model therefore sheds light on both dimensions of the evolution/rationality connection. It suggests that in a simple case, rational-like behaviour will evolve; and, as applied to organisms that possess the requisite representational states, suggests that evolution will lead them to update their beliefs, or proto-beliefs, according to Bayesian norms. It also shows why it may be useful, for the purposes of adaptationist theorizing, to treat an organism capable of flexible behaviour as if it were a rational agent, whether or not its behaviour actually stems from internal mental representations. For as long as its behaviour has been optimized by natural selection, it will behave identically to how a rational agent, who cared about its biological fitness, would behave.

Also, our model illustrates how adaptive behaviour can result from the integration of innate and acquired information. The prior probability distribution can be interpreted as information that has been hard-wired into the organism's genome by natural selection; the environmental signal then corresponds to acquired information; and the conditionalization operation represents the integration of the two sorts of information. Thus the model captures, in a precise way, the familiar point that adaptive fit between organism and environment can result from selection, learning, or a combination of the two.[13]

Finally, the model rests on a number of critical assumptions. First, an organism's payoff is assumed to depend only on its own choice, so game-theoretic interaction and frequency-dependent selection are assumed absent. Second, the model applies to a single choice, rather than a sequence of choices. Third, the payoff to a given choice is assumed to depend only on the state of the world, and not on the organism's internal state. Fourth, one particular criterion of evolutionary success—expected

[13] See McNamara and Dall (2010) for further discussion of this point.

reproductive output—was employed, which is known not to be universally applicable. These assumptions give a clue as to how models of the evolution of irrationality work, discussed in the following chapter.

6.6 Fitness and Utility

The apparent isomorphism between fitness and utility has been alluded to already. The former is the quantity that an organism's evolved behaviour will maximize, given standard adaptationist assumptions; the latter is the quantity that a rational agent's behaviour will maximize, given the assumptions of standard rational choice theory. The two quantities thus play structurally similar roles in their respective theories. This point has long been recognized; for example, it was part of Maynard Smith's motivation for importing the Nash equilibrium concept of classical game theory into biology (with modifications). Essentially, Maynard Smith's idea was to interpret the payoffs of a game as fitness increments rather than utilities, and to let natural selection replace rational deliberation as the mechanism for achieving equilibrium.[14]

In the simple Bayesian model above, we saw that by equating utility with fitness, that is, incremental number of offspring, and by assuming that the organism's probabilistic beliefs matched the objective frequencies, we could make biologically optimal behaviour coincide with rational behaviour: expected fitness corresponds to expected utility. In the next chapter we shall see that this simple equation does not always work. But before that, it is worth asking in more abstract terms how similar fitness and utility are, in respect of both their theoretical roles and the types of quantity that they are.

On the standard 'revealed preference' approach to rational choice, which corresponds to Kacelnik's E-rationality, utility is a purely notional quantity. An agent's utility function is a mathematical device by which their preferences may be described or represented. Thus in the simplest case of choice under certainty, there is a finite set of options O over which the agent has a (weak) preference R, which is assumed to be transitive and complete. The agent's utility function u is any real-valued function on the set X which is such that $u(x) \geq u(y)$ if and only if xRy, for all $x, y \in O$. Thus if the agent is rational, they behave as if they are trying to maximize their utility. But as noted earlier, there is no assumption that the agent is consciously aiming at this, and the utility function need have no psychological reality.[15]

Fitness is similar to this in one respect. Biologists who treat organisms as fitness-maximizers are obviously not suggesting that they consciously try to maximize their fitness; rather the suggestion is that natural selection has led them to behave as if they did. So the maximization is in an 'as if' sense, just as in utility theory. But there is also a disanalogy, for fitness is a real quantity, not a notional one. An organism's

[14] See Maynard Smith (1982). For a good discussion see Hammerstein (2012).

[15] Though there is an older tradition of treating utility as an actual psychological magnitude, which some philosophers still defend.

fitness, in the simplest case, is the number of offspring it produces in its lifetime; this is an actual magnitude, not one introduced to represent a binary relation. Indicative of this difference is that the conceptual priority of preference over utility is reversed for fitness. As we saw in section 1.7, talk of what an organism 'prefers' or 'wants' is commonplace in biology, but such talk is derivative from the fitness function, not used to define it. So in this respect, the situation is the opposite from in utility theory, at least on the revealed preference construal of the latter.

Another point of comparison concerns measurement scale. For utility, one standardly distinguishes between ordinal, cardinal, ratio, and absolute measurement scales, depending on the class of transformations to which an agent's utility function can be subjected without loss of information. An ordinal scale is unique up to increasing transformation; a cardinal scale up to affine (positive linear) transformation; a ratio scale up to multiplication by a positive constant; and an absolute scale only up to the identity transformation.[16] Different measurement scales are needed in different contexts. For example, cardinal utility is needed for the expected utility analysis of choice under uncertainty, but ordinal utility suffices for describing choice between certain options, for example, commodity bundles. In some applications of social choice theory, ratio-scaled utility is needed.[17]

What about biological fitness? What can be said about measurement scale here? In many evolutionary models fitness is defined on an absolute scale, for example, as expected number of offspring, or probability of survival, or intrinsic growth rate, or genetic contribution to the future population. These quantities are all absolutely measurable, in that the actual numbers are meaningful. However, it does not follow that for the purposes of evolutionary analysis, all of this information is necessary, and often it is not. For in many cases what matters is how many offspring an organism or genotype leaves relative to others in the population, as this determines whether it will be favoured by natural selection or not.

For certain evolutionary questions, a purely ordinal fitness concept suffices. For example, if competing strains of bacteria are growing in a nutrient broth, then under standard assumptions the strain with the highest intrinsic growth rate will be the only one found at equilibrium. So to predict the equilibrium state, we need only be able to order the strains by growth rate; the actual growth rates do not matter. But in other cases ordinality is not enough. For example, in a one-locus two-allele population genetics model with random mating, we know that the population will converge to a polymorphic equilibrium if the heterozygote is fitter than the homozygotes (i.e. $w_{Aa} > w_{aa}$ and $w_{Aa} > w_{AA}$); but to determine the population's equilibrium composition we need to know the fitness differences between the three genotypes, which requires cardinal measurement. And to determine the rate at which the population approaches

[16] See Roberts (1985) for a good introduction to these concepts.

[17] See for example Sen (1977b).

equilibrium, we need to know the ratios of the genotype fitnesses to one another, which requires ratio-scale measurement.

To focus the issue more generally, consider again the Price equation, used to express gene frequency change in a population over a single generation:

$$\Delta p = \frac{1}{\overline{w}} \left\{ \mathrm{Cov}(w_i, p_i) + \mathrm{Exp}(w_i \delta_i) \right\}$$

Here w_i is the fitness of the ith individual, $\overline{w}$ is average population fitness, p_i is the gene frequency of the ith individual, δ_i is the difference in gene frequency between the ith individual and its offspring, p is the population-wide frequency of the gene, Δp is the change over a single generation, and 'Cov' and 'Exp' denote covariance and expectation, respectively (Price 1970). It is easily seen that the class of transformations to which the fitness function w may be subject, while leaving Δp unchanged, involves multiplication by a positive constant, implying ratio-scale measurement. Note, though, that if we are only interested in whether the gene will spread, and if transmission bias is ignored, that is, $\delta_i = 0$ for all i, then cardinal measurement is enough; since adding a constant to the fitness function will leave the sign of Δp unchanged.

There are other contexts in which cardinality will also suffice. For example, in one-population evolutionary game theory, in which organisms are drawn at random from an infinite population to play a symmetric game, the standard replicator dynamics are unaffected by addition of a constant to each payoff, and multiplication of each payoff by a positive constant only changes the velocity with which the population moves through the state space, not the orbits (Weibull 1995, pp. 73–4). (Though other dynamics can be affected, and in two-population models the situation is different.) So for the purposes of answering the questions that evolutionary game theory usually asks, fitnesses need only be cardinally measurable. But in general, it cannot be assumed that adding a constant to the fitness function makes no difference, as the zero point of fitness is meaningful.

Finally, what about interpersonal comparison? In most rational choice analysis, utility is assumed not to be interpersonally comparable: one agent's utility cannot meaningfully be compared with another's. It might be thought that there is a dis-analogy here with fitness: surely the whole point of the fitness concept is to permit comparison between the individuals in a biological population? This is true, but not entirely to the point. For in a typical evolutionary analysis there is a single fitness function, mapping traits to fitness. (This is also true in a game-theoretic context, as long as the organisms are playing a symmetric game.) So the fact that a population contains many organisms, whose fitnesses may be compared, does not automatically constitute an analogue of interpersonal comparability. Different organisms have different traits or strategies, hence they have different fitnesses; but this is simply the analogue of a rational agent receiving different amounts of utility from different outcomes, which involves only intrapersonal comparison.

However, in some evolutionary analyses, the biological analogue of interpersonal utility comparison does arise. We encountered an example in section 2.5, in our treatment of the veil-of-ignorance concept in biology; this involved multiple individuals in a group each with a different fitness function—since the functions were defined over distributions of a resource among the group's members.[18] Since biological fitness is in principle measurable on an absolute scale, it is unproblematic to make interpersonal comparisons; thus, for example, it is perfectly meaningful to add the fitnesses of different individuals in a biological group, if we wish to do so. Here too there is a disanalogy with utility; this is precisely why the veil-of-ignorance idea applies more cleanly in a biological context, as we saw.

To summarize, in principle biological fitness is measurable on an absolute scale; but depending on the question we are trying to answer, and the details of the evolutionary model, weaker measurability assumptions will often suffice. For utility, the appropriate measurement scale also depends on the details of the model, and thus the question being addressed. However, since utility is a notional not a real quantity (on the orthodox revealed preference view), the use of (say) a cardinal measurement scale does not mean that the richer information is there but is not needed, as it does for fitness.

6.7 Naturalization of Rationality?

The attribute of rationality, in the sense of either E- or P-rationality, is an unusual organismic attribute, for it has an irreducibly normative dimension. To call an action or belief 'rational' is not just to describe it but also to evaluate it. Ideally, a full account of rationality would explain not just why organisms exhibiting the cognitive and behavioural traits that we call 'rational' evolved, but also the normative dimension of our rationality assessments. Agents who perform faulty inferences, or who fail to choose actions appropriate to their goals, or who violate canons of Bayesian rationality, are regarded as having done something *wrong*. Do these normative judgments have a biological basis?

I can see two possible answers. One is to invoke the fact/value dichotomy, and to insist that an evolutionary account is silent with respect to normative matters. Just as evolution is silent about ethical norms—it may tell us why humans exhibit certain behaviours but not whether they are right or wrong—so it is silent about rational norms, on this view. The second more ambitious answer is that the rational/irrational distinction can ultimately be naturalized in terms of the adaptive/maladaptive distinction.

[18] Another example where the biological analogue of interpersonal comparison arises is the study of symbiotic interaction, involving individuals from different species; the evolutionary analysis must then study both populations at once. The same is true in two-population evolutionary game theory more generally; see Weibull (1995, p. 174) and Hammond (2005) for useful discussion.

To motivate this second answer, consider an analogy from the philosophy of medicine. The distinction between health and disease has both a normative and descriptive dimension, like the rational/irrational distinction. 'Healthy' both picks out a certain class of physiological conditions and evaluates them positively. Beginning with Boorse (1975), many authors have tried to naturalize the health/disease distinction in terms of evolved function. The idea is that a bodily organ or subsystem is healthy if and only if it performs its function, that is, does what it is 'meant' to do in the evolutionary sense. Thus a kidney that fails to remove waste products from the blood is diseased, since removing waste is what kidneys are meant to do.

The parallel claim about rationality would be that an organism's cognitive capacities have evolved functions, which include producing behaviour that is suitably based on reasons (P-rationality) and internally consistent (E-rationality). When these capacities work as they should, the organism performs instrumentally rational actions, reasons well, chooses consistently, and satisfies Bayesian principles. When the capacities malfunction, the organism ends up violating these rationality criteria. Thus the rational/irrational distinction, like the health/disease distinction, is naturalized in terms of biological function.

This is a coherent suggestion philosophically, but it faces an empirical challenge from the experimental evidence on violation of rationality norms, discussed above. For the suggested naturalization to succeed, this evidence would have to be reconciled with the claim that the organisms in question, human and non-human, have evolved capacities whose function it is to satisfy the norms of E- and P-rationality, even though they often fail to do so. In principle this could be done, for example, by arguing that the experimental settings in which the anomalies above are found differ from the relevant evolutionary environments, so are not a good guide to evolved function; or, in the human case, that social and cultural factors are serving to mask our evolved psychological dispositions. After all, it is a familiar point that an organ will not always perform its evolved function if it is placed in a novel environment, or if other distorting factors are present.

I think this strategy is ultimately unlikely to succeed. There is certainly an issue about whether the experimental set-ups are suitably naturalistic, particularly for animals, and a broader issue about how best to interpret the observed patterns found in these experiments. However, the violations of E- and P-rationality are both widespread and systematic, and in the human case found across cultures; they do not seem much like performance errors. So the default assumption should surely be that, for whatever reason, subjects have an evolved predisposition to violate these rationality criteria, rather than to satisfy them if only extraneous factors had not intervened. This is particularly the case if it turns out that there are adaptive explanations for why people exhibit the irrationalities of choice that they do; see chapter 7.

This suggests that the norms of E- and P-rationality are unlikely to succumb to evolutionary naturalization. To see where this leaves us, let us return to the medical analogy. Suppose it turns out that the health/disease distinction, as actually used in medical practice, does not coincide with the function/malfunction distinction, so cannot be naturalized in the desired way. Two options would then be available. The first is to seek a different, non-naturalistic account of the normative significance of the health/disease distinction, for example, by accepting that it involves society's value judgments. The second is to argue that the health/disease distinction should be replaced by the function/malfunction distinction, that is, to reject outright the normative significance of the former, in so far as it fails to coincide with the latter.

In the rationality case, I suggest that the first of these options is more plausible. We can grant that an agent should obey the norms of E- and P-rationality, even if their evolved psychology predisposes them not to do so in some circumstances. Fairly compelling non-evolutionary justifications exist for why an agent should satisfy norms such as probabilistic consistency, Bayesian updating, expected utility maximization, and time-consistency (though there is debate on all these points).[19] These justifications remain compelling whether or not the norms in question have an underlying biological rationale, and indeed even if there is a biological explanation for why agents systematically violate them. Thus we should allow that E- and P-rationality are *sui generis* forms of rational assessment, which cannot be reduced to biological adaptiveness.

The second, more radical option is endorsed by Gigerenzer, who is disparaging of probability and rational choice theory, which he regards as a priori exercises that do not help us understand real-life decision-making and cognition. He argues that an agent who relies on heuristics will often outperform one who tries to conform to probabilistic or decision-theoretic ideals, at least in the particular environments for which the heuristics were tailored. Moreover, Gigerenzer rejects probability and decision theory even as a normative ideal. Thus Gigerenzer and Selten (2001) say that their theory 'provides an alternative to current norms, not an account that accepts current norms and studies when humans deviate from these norms' (p. 6). Their suggestion is that the traditional norms of E- and P-rationality constitute an inappropriate standard by which to judge creatures which have evolved to be ecologically rational.

I suggest that this is an overreaction. Adaptive heuristics and ecological rationality are useful concepts, and may well point to a more descriptively accurate model of decision-making than traditional rational choice theory. But they do not force us to replace traditional rationality norms altogether; they only show that those norms cannot be naturalized. So we should accept that rationality in the sense of having good reasons for one's beliefs and actions, and in the sense of making consistent choices, are both valid forms of normative assessment, while also allowing that our cognition

19 See Pettigrew (2016) for a recent treatment of some of these justifications.

and behaviour can be evaluated in terms of ecological/biological success. These are simply different dimensions of normative assessment, which do not necessarily coincide.

6.8 Conclusion

The key to understanding the evolution–rationality connection is to recognize its two dimensions: the evolution of rationality itself, and the use of rationality-inspired heuristics in evolutionary analysis, as for example in agential thinking (type 1). The idea of adaptive behaviour as proto-rational links the two dimensions, suggesting a route by which rationality may have evolved, and an explanation for why rational idioms are well-suited to characterizing the adaptive behaviour of non-human organisms. Rationality may be understood either as consistency of choice or as having good reasons for beliefs/actions; these notions have distinct cognitive prerequisites and domains of application. The adaptive significance of rationality over arationality seems fairly clear, namely facilitating adaptive behaviour in the face of environmental complexity; but it is less clear whether natural selection would always favour rationality over irrationality. A simple model, based on optimal behaviour in an uncertain environment, suggests an evolutionary basis for Bayesian rationality; but it rests on simplifying assumptions that limit its generality. Thus despite the conceptual connection between utility and fitness, an evolutionary naturalization of the norms of rationality seems unlikely, as it would require strong empirical assumptions.

7

Can Adaptiveness and Rationality Part Ways?

7.1 Introduction

The previous chapter examined, in a preliminary way, the idea that natural selection would favour organisms whose behaviour satisfies traditional definitions of rationality. We sketched a simple illustrative model purporting to show how the norms of Bayesian rationality can be derived from a biological basis; however, we noted that the model rests on restrictive assumptions, in particular the absence of game-theoretic interaction. This chapter continues our exploration of the same theme from the reverse angle. Our focus is on a range of arguments from the literature suggesting that what is biologically adaptive and what is rational, in the sense of utility-maximizing, may sometimes part ways.[1] The notion of rationality at work in this chapter is thus E-rationality, in Kacelnik's terms.

These parting-of-ways arguments are of two sorts. The first sort suggest that natural selection will sometimes favour behaviours that cannot be characterized as utility-maximizing at all, such as, intransitive choices, or which violate more specific rationality criteria, for example, expected utility maximization. Presuming we agree that these criteria are normatively compelling, these then are arguments for the evolution of irrationality. The second sort suggest that natural selection will sometimes lead agents to attach utility to things other than their own biological fitness. These arguments do not suggest that evolution will lead to irrationality per se, but rather to the pursuit of goals distinct from fitness-maximization (at least for a particular way of defining fitness). Though seemingly different, I include arguments of both sorts under the parting-of-ways banner. For it is not always obvious which side a particular argument falls on; and those of the first sort can sometimes be reduced to the second by suitable choice of utility function and/or fitness measure.

In chapter 6, we saw that the evolution–rationality connection has two dimensions. The first is the evolution of rationality itself, considered as a phenotypic attribute that some organisms, including humans, possess to varying degrees. The second is the use of rationality-inspired concepts to understand adaptation, as in the

[1] I borrow the expression 'part ways' from Skyrms (1995).

organism-as-rational-agent heuristic—what we previously dubbed agential thinking (type 1) as applied to behaviour. Most of the arguments below are explicitly addressed to the first dimension—they offer putative explanations for why behaviours that violate traditional rationality norms might nonetheless be adaptive. However, they are also relevant to the second dimension. For if it is true that there is sometimes an adaptive advantage to irrationality, this implies a limit on the heuristic value of treating an adapted organism as akin to a rational agent trying to achieve a goal, at least if 'rational' is defined as conformity to the norms in question.

This line of argument should be distinguished from the point stressed in chapters 3 and 4, that natural selection does not necessarily lead to adaptation at all. The point in this chapter is different, namely that even if an organism's behaviour *is* well-adapted, hence furthers the organism's biological goal, it does not necessarily follow that the behaviour will count as rational, where the latter is defined in independent, and sufficiently precise, terms.

The structure of the chapter is as follows. Sections 7.2 to 7.7 consider a series of six parting-of-ways arguments drawn from the philosophy, biology, and economics literatures. I explore the logic of these arguments, explain the workings of the formal models on which they rest, and try to assess them. I consider what each argument implies for the two dimensions of the evolution–rationality connection, and ask whether the behaviours in question should be regarded as really, or only apparently, irrational. Section 7.8 steps back and tries to extract some general morals. Section 7.9 concludes.

7.2 Cooperation and the Prisoner's Dilemma

Our first parting-of-ways argument concerns the evolution of cooperation; it is found in slightly different forms in Skyrms (1995) and Sober (1998). Consider a simple Prisoner's Dilemma game. There are two choices, cooperate (*C*) and defect (*D*), with payoffs shown in Table 7.1. Suppose first that we are in a rational choice setting, in which the players are rational agents and the payoffs denote (cardinal) utilities. In the single-shot game, presuming that players cannot influence each other, it is widely agreed that the rational thing to do is to play *D*, as it strictly dominates *C*. Thus the expected utility of playing *D* must exceed that of *C*. This is so even if the agent believes that its opponent is likely to play the same strategy as itself, presuming the truth of 'causal decision theory' (Lewis 1981), as the two players are causally isolated.

Suppose we now transpose to an evolutionary setting and consider a large population of organisms, with hard-wired strategies, engaged in pairwise interaction; the payoffs now represent increments of Darwinian fitness that an organism gets depending on its own strategy and its partner's. Which strategy has the higher fitness? As Skyrms observes, this depends on the pairing assumption that we make. Under random pairing, in which the probability of having a *C* partner is same for both types, it is clear that strategy *D* must be fitter. The expected fitnesses of the strategies are:

Table 7.1. Prisoner's Dilemma

		Player 2	
		C	D
Player 1	C	(x, x)	$(0, y)$
	D	$(y, 0)$	(z, z)

Payoffs for (Player 1, Player 2), where $y > x > z > 0$

$$w_C = x \cdot \Pr(C) + 0 \cdot \Pr(D)$$
$$w_D = y \cdot \Pr(C) + z \cdot \Pr(D)$$

where $\Pr(C)$ and $\Pr(D)$ denote the probabilities of having a C and D partner, respectively; these are given by the overall frequencies of the strategies in the population. As Skyrms notes, these expressions for expected fitness are identical to the corresponding expressions for expected utility in the rational choice context, calculated using standard (Savage-style) decision theory. Under random pairing, the type with the highest expected fitness chooses the action that confers the highest expected utility, so the fitness-maximizing strategy is identical to the rational strategy.

Matters are different if there is assortment, or correlated pairing. We must then calculate the expected fitness of each strategy using conditional rather than unconditional probabilities. The resulting expressions are:

$$w_C = x \cdot \Pr(C|C) + 0 \cdot \Pr(D|C)$$
$$w_D = y \cdot \Pr(C|D) + z \cdot \Pr(D|D)$$

where $\Pr(X|Y)$ denotes the probability that an organism with strategy Y has a partner with with strategy X, where $X, Y \in \{C, D\}$. It is easy to see that if the correlation is strong enough, that is, if $\Pr(C|C)$ is sufficiently greater than $\Pr(C|D)$, then the C strategy may be fitter overall, and so spread by natural selection. Skyrms observes that these expressions for expected fitness correspond to how Jeffrey (1990) calculates expected utility in his 'evidential decision theory'—which enjoins a rational agent to take account of possible correlations, whether causal or not, between her own choice and the state of the world. Since Jeffrey's theory recommends cooperating in the one-shot Prisoner's Dilemma, Skyrms regards it as an inappropriate normative theory (as do most philosophers). He concludes that with correlated pairing, 'rational choice theory completely parts ways with evolutionary theory. Strategies that are ruled out by every theory of rational choice can flourish under favourable conditions of correlation' (1995, p. 106).

Sober (1998) develops the same point slightly differently, in the context of discussing what he calls the 'heuristic of personification' in evolutionary biology. This heuristic is the idea that 'if natural selection controls which of traits $T, A_1 \ldots A_n$ evolves in a given population, then T will evolve, rather than the alternatives, if and only if a rational agent who wanted to maximize fitness would choose T over $A_1 \ldots A_n$' (p. 409). Sober maintains that this heuristic is usually unproblematic but fails in certain contexts, one of which is the one-shot Prisoner's Dilemma. The rational agent will never play cooperate, since it is strictly dominated, Sober reasons; however, it is possible that natural selection will favour cooperate over defect if the requisite correlation exists. Thus the heuristic of personification fails: the rational strategy and the evolutionarily optimal strategy do not coincide.

Though formally identical, Skyrms' and Sober's arguments have different dialectical foci, corresponding to the two dimensions of the evolutionary-rationality connection from chapter 6. Skyrms' suggestion is that evolution may lead to strategies that are irrational, while Sober's is that a particular way of theorizing about evolution, which uses a rationality heuristic to understand natural selection, may be misleading.

Both arguments are intriguing, but there is an obvious response to each. To Skyrms' version, it can be replied that his argument does not show that irrational strategies will evolve, but rather that evolution will lead agents to have 'other-regarding preferences', that is, to have utility functions which do not depend solely on their personal gain but also on their partner's. There is nothing irrational about having such preferences, as is widely accepted, and a rational agent who has them will often behave cooperatively in social encounters. Skyrms' argument only suggests that irrationality (or the use of Jeffrey's decision rule) will evolve *if* we assume that a rational agent is purely self-regarding; but there is no particular reason to assume that.

Martens (2017) makes a nice version of this point for the special case of additive payoffs (i.e. $y - x = z$ in Table 7.1). He shows that the condition for evolution to favour strategy C, that is, $w_C > w_D$, will be identical to the condition for a rational agent who uses the standard Savage decision-rule (i.e. causal decision theory) to choose C over D, as long as the agent's utility function depends in a certain way on their partner's payoff as well as their own. (Their utility should equal inclusive rather than personal fitness.) Thus while Skyrms presents his parting-of-ways argument as one of the first sort (evolution $\Longrightarrow$ irrationality), it can also be interpreted as the second (utility $\neq$ fitness). I suggest that the latter interpretation is more reasonable, since there is evidence that humans have other-regarding preferences, but there is little evidence (so far as I am aware) that they knowingly conflate correlation with causation when deciding how to act.

To Sober's version, it can be replied that his heuristic of personification relies on a specific definition of fitness, namely personal reproductive output, which is not necessarily appropriate in a social context. It is a familiar point in evolutionary biology that where social encounters are between kin, or more generally between organisms of the same genetic type, evolution may favour 'altruistic' traits, which do not maximize personal fitness. As we saw in chapter 5, part of the point of the 'inclusive fitness' concept of Hamilton (1964) was precisely to supply a fitness definition for use in a social context, relative to which the heuristic that Sober describes would work. Indeed Hamilton (1971) himself noted that in a one-shot Prisoner's Dilemma played between relatives, if the payoffs are taken to be inclusive rather than personal fitness then cooperation may be the dominant strategy—thus restoring the link between evolutionary optimality and rationality. We saw in chapter 5 that this argument is valid, but relies essentially on payoff additivity (an assumption that Sober (1998) also makes). With non-additivity matters are different; however, there is still a way of restoring the evolution–rationality link and the personification heuristic by suitable transformation of the payoff matrix, as we saw in section 5.4.

To conclude, the Skyrms/Sober point is an interesting one; however, it need not be taken as showing that evolution will lead to irrationality, nor that the organism-as-rational agent idea fails in social contexts. These conclusions only follow if we make specific, and non-mandatory, assumptions about what rational agents care about and/or what definition of fitness to use.

7.3 Fairness and the Ultimatum Game

Our second parting-of-ways argument concerns strategic behaviour in the ultimatum game; versions of the argument are found in Skyrms (1995), Samuelson (1997), and Rand *et al.* (2013), among others.

The ultimatum game, or 'divide-the-dollar', is a simple two-player game much studied in experimental economics. Player 1 (the proposer) is given a dollar and must propose a way of splitting it with player 2 (the responder), who has two choices: accept, in which case both players receive their proposed share, or reject, in which case both players get nothing. What is the rational thing to do? One familiar argument says that the responder should accept whatever they are offered, since something is better than nothing; and the proposer, anticipating this, should offer the minimum possible share, say one cent, keeping ninety-nine cents for themselves. However, this is not what happens empirically. Rather, responders tend to reject manifestly unfair offers, even in one-shot games; and proposers tend not to make them. This is a well-confirmed empirical result.[2]

The strategy pair in which the proposer offers the minimum and the responder accepts any offer is a Nash equilibrium of the ultimatum game—neither party gains by unilateral deviation. However, there are many other Nash equilibria too. For example, if the proposer offers forty cents and the responder accepts offers of at least forty cents but rejects others, the resulting strategy pair is also a Nash equilibrium.[3] However, this Nash equilibrium has an undesirable feature, as the responder's strategy calls on them to 'leave money on the table' in the event of receiving too unfair an offer, which is intuitively irrational. This Nash equilibrium is thus not 'sub-game perfect', since it is not an equilibrium of the sub-game that starts from the point at which it is the

[2] See Camerer (2003) and Güth and Kocher (2013) for surveys of the empirical results.

[3] More generally, if the proposer offers p cents, and the responder accepts any offer of at least p cents, then Nash equilibrium results.

responder's turn to move, the proposer having unfairly offered less than forty cents. The only sub-game perfect Nash equilibrium of the ultimatum game is the one where the proposer offers the minimum and the responder accepts everything.

To see why the responder strategy of rejecting an unfair offer is irrational, note that it is weakly dominated by the strategy of accepting all offers: the latter never gets a lower payoff and sometimes gets a higher one.[4] Though less obviously than for strictly dominated strategies, there is something anomalous about a player who wishes to maximize their payoff employing a weakly dominated strategy. Note also that if the proposer believes that the responder will not use a weakly dominated strategy, then their best choice is to offer the minimum. So in short, if both players are rational and know that the other is rational, then the sub-game perfect equilibrium should result; but empirically it does not.

One argument says that observed behaviour in the ultimatum game is not actually irrational, since players' utility functions need not depend on their monetary payoff alone. Rather, players are partly motivated by a hard-wired sense of fairness; this leads responders to reject manifestly unfair offers and proposers not to make them. Other researchers suggest that 'spite', and 'empathy' are affecting players' choices.[5] Such psychological factors may well be part of the proximate explanation for why people behave as they do in the ultimatum game, and offer a possible way of reconciling the data with the utility-maximizing paradigm. But as has often been noted, there is a deeper question about how a concern for fairness could have evolved in the first place. Could proposers who make fair offers, and responders who reject unfair ones, survive the evolutionary process?

This question has heightened significance given that there are biological scenarios with a structure similar to that of the ultimatum game. For example, consider the conflict of interest between a male and a female in a breeding pair over who will provision for their offspring, often modelled as a 'mate-desertion game'.[6] In this game, both the male and female have two options: stay and care, or desert. Suppose that the male chooses first. If the male deserts, the female's best choice is to stay and care; but if the male stays and cares, the female should desert; payoffs here are in increments of biological fitness (Figure 7.1). There is a unique sub-game perfect equilibrium in which the male deserts and the female uses the strategy 'stay if male deserts; desert if male stays'. However, there is another equilibrium in which the male stays and the female uses the strategy 'always desert'. Again, this is not sub-game perfect; since the female's strategy calls on her to make an inferior choice in the event that the male were to desert.

[4] A related point is that the responder strategy of rejecting unfair offers violates 'sequential rationality' (Kreps and Wilson 1982), also known as 'modular rationality' (Skyrms 1995), which requires a player to make a payoff maximizing choice at every conceivable point in a game.

[5] See Debove *et al.* (2016) for a recent survey of the possible explanations.

[6] Maynard Smith (1977) was the first to analyse parental investment in game-theoretic terms.

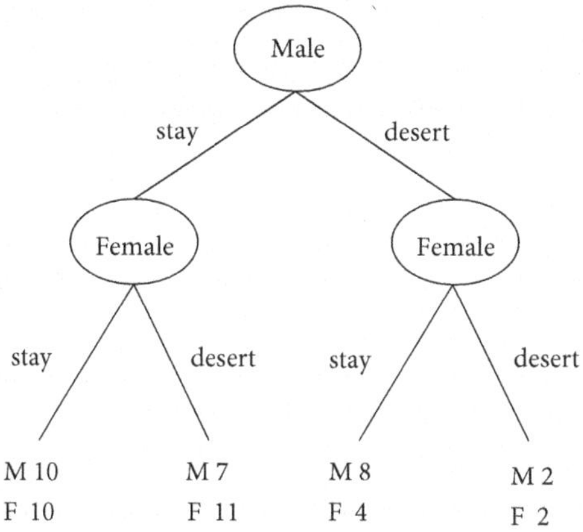

Figure 7.1. Desertion game

From an evolutionary perspective, intuitively one might expect natural selection to produce the sub-game perfect equilibrium in the ultimatum game. For consider an alternative equilibrium in which proposers all make relatively fair offers and responders reject highly unfair ones. Such an equilibrium seems unstable, hence biologically unrealistic. For if proposers occasionally make unfair offers by mistake, or if mutation introduces a handful of proposers with the unfair strategy, then a responder who accepts all offers will have an advantage. If such responders spread, this in turn gives the advantage to proposers who make unfair offers. Thus the system will evolve towards the sub-game perfect equilibrium in which proposers offer the minimum and responders accept everything. This argument was originally used by Selten (1975) to justify sub-game perfection and is often appealed to in a biological context.[7]

However, explicit modelling of the evolutionary dynamics has cast doubt on this intuitive argument.[8] In the standard treatment, agents are randomly matched to play the ultimatum game using a hard-wired strategy—which specifies what to offer as responder, and what to accept as responder. Evolutionary competition between the strategies is then modelled by the replicator dynamics—in which a strategy's spread or decline depends on whether its payoff is greater or lower than the average payoff. Two key results emerge. First, the population does not necessarily evolve to the sub-game perfect equilibrium. Rather, the system may converge on an equilibrium in which some proposers make fair offers and some responders reject unfair ones. Second, such an equilibrium need not be unstable in the presence of mutational noise; it depends on the details of how the mistakes are made.

What explains these results? The first occurs because the replicator dynamics will not necessarily eliminate weakly dominated strategies (Samuleson 1988,

[7] See for example Hammerstein (2012) and McNamara and Houston (2002).
[8] See in particular Gale *et al.* (1995), Skyrms (1995), Samuelson (1997), and Rand *et al.* (2013).

Skyrms 1995). The responder strategy in which unfair offers are rejected may seem irrational, but will only be penalized if proposers who make unfair offers are actually present in the population. If such proposers are driven to a low enough frequency, responders who reject unfair offers can persist. The second result occurs because mistakes, or mutations, continually introduce new responder strategies into the population. If this happens at a sufficient rate, the resulting spread of responder strategies, some of which will reject unfair offers, can stabilize the equilibrium in which proposers make fair offers.

What does all this show? There are two morals, corresponding to the two dimensions of the evolution–rationality connection. The first is that under favourable conditions, evolution may lead to behaviour similar to that seen in ultimatum game experiments; this is the moral standardly drawn in the literature. If one is persuaded that such behaviour is irrational, this can be interpreted as the evolution of irrationality. However, as with the Prisoner's Dilemma case, it could instead be taken to show that biological evolution will equip agents with utility functions that depend on more than their personal payoff or fitness, for example, that a preference for fair division will evolve. In so far as the aim is to to account for the human experimental data, I see no decisive reason to favour the 'evolution $\Longrightarrow$ irrationality' over the 'utility $\neq$ fitness' interpretations; the choice depends mainly on whether one cleaves to a strict revealed-preference view of utility-maximization or not.

However, as regards the second dimension of our evolution–rationality connection, I favour the former interpretation. For the trick of preserving rationality by redefining the utility function threatens to obscure the true lesson that the evolutionary game-theoretic analysis has for the organism-as-rational-agent heuristic, which goes beyond the ultimatum game per se. The lesson is this: if we wish to treat evolved organisms as akin to rational agents, using strategies that further their biological goals, then we need to tread with care. For a plausible assumption about rational agents is that they will never use weakly dominated strategies—however, Darwinian evolution, surprisingly, may favour organisms that do. A biologist who observes an evolved organism using such a strategy, for example, a female who uses 'always desert' in the mate-desertion game of Figure 7.1, may be tempted to dismiss it as mal-adapted, given its apparent irrationality; but this does not necessarily follow. Reliance on a rationality heuristic can lead astray.

The practical significance of this point is unclear. Hammerstein (2012) states that there is a 'plethora of biological examples' in which sub-game perfect equilibrium is found—which suggests that non-humans approximate the rationality ideal of eschewing weakly dominated strategies better than humans do (p. 18). On the other hand, chimpanzees have been reported to play the ultimatum game similarly to humans, that is, by rejecting offers that are too unfair (Proctor *et al.* 2013). Here is not the place to adjudicate this empirical matter. Rather the point is that in principle, the case of sub-game perfection shows one way in which the isomorphism between the rational and the biologically optimal can break down.

7.4 Trust and the Indirect Evolutionary Approach

Our third parting-of-ways argument comes from W. Güth's 'indirect evolutionary approach'.[9] This denotes a specific way of combining evolution and rational deliberation into a single model. By contrast with evolutionary game theory, which evaluates competing strategies in terms of their objective payoff (fitness), and does not explicitly consider proximate motivations, the indirect approach deals with objective payoffs and agents' subjective utilities (or preferences) at the same time. The basic idea is that how an agent plays a game in real time will depend on their subjective utilities over the possible outcomes; but over evolutionary time, success depends on the objective payoffs. Thus at each time, agents choose in accordance with their subjective preferences in a rational manner; but over time, preferences evolve. This enables Güth to pose an interesting question: will evolution always favour agents whose subjective utilities are equal to (or linear in) the objective payoffs?

At first glance, the answer might seem to be obviously 'yes'. Surely evolution can do no better than build an agent whose proximate motivation is to maximize its objective payoff, since the latter is what determines evolutionary success? However, Güth is able to show that this need not be true; it all depends on whether agents' preferences are observable or not. To illustrate, consider the simple two-player 'trust game' in Figure 7.2 played between rational agents. Player 1 chooses first, and can either defect (D), in which case the game ends with payoffs of $(0, 0)$, or place trust in her opponent (T). In case of the latter, player 2 can then choose to reward the trust (R), leading to payoffs of $(1, 1)$, or exploit the trust (E), leading to payoffs of $(-1, b)$, where $b > 1$. These are the objective payoffs, hence determine the evolutionary success of

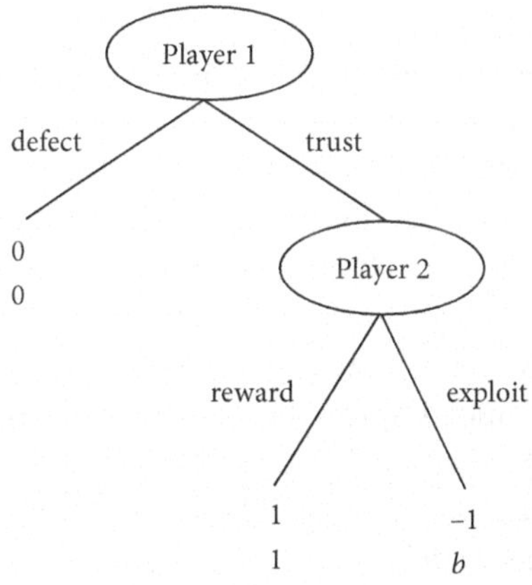

Figure 7.2. Trust game
Source: modified from Berninghaus *et al.* (2012), p. 114, by permission of Cambridge University Press

[9] See Güth and Yaari (1992), Güth and Kliemt (1998), and Güth and Peleg (2001).

the strategies. There is a unique sub-game perfect equilibrium in objective payoffs, in which both players use the strategy 'D in the player 1 role, E in the player 2 role', leading to payoffs of $(0, 0)$.

Güth now argues as follows. Suppose first that all agents' subjective utilities coincide with the objective payoffs. Since the agents are rational, they will not use dominated strategies, so the sub-game perfect equilibrium will result. Next, consider a trustworthy agent who is genuinely moral, so dislikes defecting when player 1 has reposed trust in them. Such an agent's utility function can be characterized by replacing b with $b^* < 1$, leaving other payoffs unchanged; the agent will now be motivated to play R when in the player 2 role. Conceivably, the trustworthy b^* type may earn a higher objective payoff than the b type, despite not being subjectively motivated to maximize it. For if player 1 can gain information (not necessarily perfect) about whether their opponent is trustworthy, that is, is of type b or b^*, they can condition their choice on this information. In the limiting case of perfect information about opponent's type, player 1 will play T if and only if their opponent is of type b^*. Then, the b^* type will do better (objectively) than the b type, as the latter will never have a chance to exploit player 1's trust; so evolution will lead to a population of trusting, trustworthy agents, that is, who play T in the player 1 role and R in the player 2 role. More generally, if imperfect information about type is available, even at some cost, trust may still evolve.[10]

This is an intriguing argument, but what exactly does it show? Note first that while Güth frames the argument as a parting-of-ways of the 'utility $\neq$ fitness' sort, it could also be interpreted as the 'evolution $\Longrightarrow$ irrationality' sort; the issues here are identical to those in the previous section. Second, critics have pointed out that what drives Güth's argument is really the assumption about the observability of player's types (preferences), rather than the separation between subjective and objective payoff per se (Robson and Samuelson 2010). If type is completely unobservable, trust will not evolve in the model above.[11] Moreover, treating a player's type (or preferences) as observable gives rise to further problems. One cannot simply assume that trustworthy b^* types are able to signal this fact, for it will clearly pay an untrustworthy b type to fake the signal, to induce their opponent to trust them. One might think it would pay agents to invest in the technology needed to reliably detect their opponents' type; this is true, but it is equally advantageous to conceal one's type from one's opponent. So an arms race may ensue, in which improved detection leads to improved concealment, and so on (Robson and Samuelson 2010, McNamara and Houston 2002). It seems unlikely that anything general can be said about how such an arms race will end, at least on the basis of simple models.

[10] The case of imperfect information about type is analysed by Güth and Kliemt (1998) and Dekel *et al.* (2007).

[11] Though in more complex models, evolution can lead to the sub-game imperfect equilibrium, as we have seen in section 7.3.

To conclude, the indirect evolutionary approach offers a useful take on the evolution–rationality connection (first dimension). Its conceptual separation of rational deliberation from preference evolution is more realistic than some ways of applying evolutionary theory to human behaviour, and offers a precise way of formulating the parting-of-ways hypothesis. Moreover, the main moral—that if type is at least partially observable, evolution may favour agents who are subjectively motivated not to maximize objective payoffs—is an instructive one. However, given the difficulty of explaining how this observability arises and persists, it is unclear whether the resulting parting-of-ways is a robust phenomenon or not.

Finally, what are the implications for the second dimension—the organism-as-rational-agent heuristic in biology? One might think that Güth's results pose a challenge for this heuristic. If evolution can favour agents who are not motivated to maximize fitness, surely treating evolved organisms as if they were rational agents with that goal looks problematic? However, I think the correct conclusion is that care is needed. For what drives Güth's results is really just the fact that in a strategic context, what players know about each others' types may make a difference to the rational choice of action. This point applies in biology too, with 'optimal' replacing 'rational', and plays a central role in the literature on mimicry, deception, and signalling.[12] It is an instance of the general point that to determine whether an evolved behaviour is optimal, it is crucial to take account of the information that the organism has at their disposal. So the indirect evolutionary approach does not undermine the agential heuristic, so much as draw attention to something that biologists who use the heuristic already know.

7.5 Intransitive Choices

Our fourth parting-of-ways argument concerns individual decision-making, with no game-theoretic component. That an agent's preferences should be transitive, that is, that if they prefer *A* to *B* and *B* to *C* they should prefer *A* to *C*, is a fundamental aspect of rationality, and normatively very compelling (though this has occasionally been questioned).[13] An agent whose preferences violate transitivity cannot be represented as maximizing a utility function; so transitivity is built into the fabric of the utility-maximizing paradigm of rational choice theory. However, apparent violations of transitivity have been reported empirically, in both humans and animals.[14]

On the face of it, transitivity seems like an implausible candidate for a parting-of-ways argument. If an evolved organism exhibits adaptive behaviour, then it seems obvious that its preferences or choices must be transitive—it should always choose

[12] See for example Maynard Smith and Harper (2003).

[13] Authors questioning transitivity include Schumm (1987) and Fishburn (1991).

[14] Shafir (1994) and Shafir *et al.* (2002) find intransitive choices in honey bees. Kalenscher and Pennartz (2011) review the findings on intransitive choices in humans.

the option that brings higher fitness. However, recent work by Houston *et al.* (2007b) and McNamara *et al.* (2014) suggests that matters are not so simple. These authors construct models in which evolutionary optimality requires that an animal actually make intransitive choices. I briefly explain how such a counterintuitive result emerges, then ask what it teaches us about the evolution–rationality connection.

In the first model of Houston *et al.* (2007b), an animal must make a foraging choice at each of a large number of discrete times, $t = 0, 1, 2 \ldots$.[15] There are three options to choose between, *A*, *B*, and *C*, but only two are available in any particular environment. Each option has a different probability of yielding a food item with a given energy content (common to all three). The options also differ in terms of the predation risk incurred. Option *A* is safest but has the lowest chance of success; option *C* has the highest chance of success but is riskier. Option *B* brings intermediate success and incurs the same predation risk as *C* (Table 7.2).

The animal's 'goal' is to survive the winter; to achieve this it must avoid starvation, which occurs when its energy reserves fall below a certain level, and also avoid being killed by a predator. The animal's strategy specifies what option it should choose in each environment (or choice situation), as a function of its current energy reserves. The optimal strategy is the one that gives the highest probability of surviving the winter; under standard adaptationist assumptions, this is what we expect natural selection to favour. Houston *et al.* (2007b) find that for a range of intermediate energy reserves, the optimal strategy requires the animal to make intransitive choices: it should prefer *C* to *B*, *B* to *A*, and yet *A* to *C* (Table 7.3).

Table 7.2. Foraging options

Option	Probability of success	Predation risk
A	Low	Zero
B	Medium	$\mu > 0$
C	High	$\mu > 0$

Source: based on Houston *et al.* (2007b), p. 366, Table 1

Table 7.3. Choices that maximize survival

	A versus *B*	*B* versus *C*	*A* versus *C*
Low reserves	$B > A$	$C > B$	$C > A$
Intermediate reserves	$B > A$	$C > B$	$A > C$
High reserves	$A > B$	$C > B$	$A > C$

Source: based on Houston *et al.* (2007b), p. 366, Figure 1

[15] For expository purposes I focus just on this model; however, the morals I draw apply to the other models in Houston *et al.* (2007b) and McNamara *et al.* (2014) too.

What explains this result? The second column is straightforward: *C* is always a better choice than *B*. The key lies in the first and third columns. When the animals' energy reserves are low, it should favour the riskier options (*B* and *C*) over the safer *A*—as the need to avoid starvation outweighs the predation risk. Conversely when reserves are high, the safer option (*A*) is preferable. So as we move up the table, from high to low reserves, a switch in favour of the riskier option occurs in both the first and third columns. However, the energy level at which the switch occurs differs between the columns. When the options are *A* and *B* (column 1), the switch occurs at intermediate reserve levels; but when they are *A* and *C* (column 3), it does not. The reason is that *C* is more profitable than *B*; so in an environment where *C* is available, the animal has better insurance against future starvation. It can thus afford to wait until its reserves levels are low before eschewing the safe option.

In effect, the intransitivity in this model arises because the animal's probability of surviving the winter depends on its repeated choices; and because current options are presumed also to be available in the future, that is, the environment stays stable. The animal's energy reserves may vary over time, so natural selection equips the animal not with a fixed preference between options, but rather with a strategy which specifies which option to choose as a function of its energy reserves. At any one time, the options available provide information about which environment the animal is in—and thus about which options will be available in the future. Since survival probability depends on future choices as well as current choice, the optimal strategy must take account of this information. This is what accounts for the preference reversal occurring at different points in columns 1 and 3, which leads to the intransitivity.

Importantly, there is no intransitivity at the level of strategies (as Houston *et al.* (2007b) note). Strategies can be linearly ordered by their fitness (survival probability). So it is not that natural selection's 'preference' between strategies is somehow irrational. Rather, it is the animal's choices at a given time that are intransitive (at intermediate reserve levels), but this is because they are a one-time manifestation of a strategy designed for repeated use. So in a way, the anomaly dissolves once we apply the fitness concept at the level of strategies. It is natural to try to attach a fitness value to each foraging option itself (*A*, *B*, or *C*) in the above model—in which case intransitive choice would clearly conflict with fitness-maximization. But this would only be appropriate if evolution had equipped the animal with a fixed preference between the options. If instead evolution has equipped the animal with a reserve-dependent strategy, it is strategies not choices that should be evaluated by their fitness.

The general moral here is that a behavioural choice taken in isolation may seem anomalous, but makes sense if viewed as resulting from an evolved strategy. A similar point is often stressed in the behavioural economics literature, namely that apparent anomalies of choice, such as cooperation in a social dilemma, may arise because experimental subjects rely on strategies that have evolved for use in repeated games, where reputation effects, for instance, play a role. Such strategies may well be adaptive, but generate anomalous results in single-shot experiments. Similarly, the experiments that find intransitive choices among animals may be eliciting behaviour that stems from a strategy designed for making repeated choices; as the model above shows, such a strategy may be optimal.

What then does this parting-of-ways argument imply? Does it show that evolution will sometimes favour irrationality? The answer seems to be yes. However, once we understand how the model works, it may seem tempting to dismiss the irrationality as only apparent—for it involves comparing choices made in different circumstances. An animal facing a choice between A and B is in a different environment to one facing a choice between B and C, and this makes a key difference. The former choice is not between A and B *simpliciter* but rather between A and B *when both are available and will continue to be.* This move reconciles rationality with evolutionary optimality; the downside is that it renders the transitivity principle immune from empirical test in the above model, since every pair of choices defines a distinct environment.

Interestingly, a parallel issue arises in a non-biological context, in discussions of the normative status of the transitivity axiom. In an example from Broome (1991), a person prefers visiting Rome (R) to a mountaineering trip in the Alps (M), prefers staying at home (H) to the Rome visit, but does not prefer staying at home to the mountaineering trip—thus violating transitivity (p. 100). The person rationalizes their preferences as follows. They find mountaineering scary, so $R > M$; they find museums boring, so $H > R$, but they regard staying at home instead of going mountaineering as cowardly, so $M > H$. But the intransitivity can be eliminated by splitting H into two: staying at home having declined a mountaineering trip (H_1); and staying at home without having done so (H_2). This has a certain plausibility; but as Broome notes, unless some limit is placed on the permissible re-descriptions the transitivity requirement will simply become empty.

There are murky philosophical issues on the horizon here (about the identity and distinctness of options) that I cannot get into. I incline towards the view that the re-description strategy is not always permissible, and thus that at least some of the transitivity violations reported among humans do constitute real irrationality, whatever their psychological basis. Similarly, I am inclined to take models for the evolution of intransitive choices at face value, as showing that irrationality can, in specific environmental contexts, be biologically optimal. However, I have no knock-down argument against those who would interpret the irrationality here as merely apparent.

Finally, what of the organism-as-rational-agent heuristic? Does the above model, and others like it, threaten its coherence? I think not. Rather they reinforce our previous moral: the heuristic must be applied with care. In previous partings-of-ways cases, we found that the utility function needed to be carefully defined to make the heuristic work. In this case, it is not the utility function but rather the state-space of the decision problem that must be chosen appropriately. If we envisage an evolved organism as choosing not between options but rather between strategies, with the goal of maximizing its survival prospects, then the organism-as-rational-agent idea works fine. The animal does indeed behave like a rational agent whose goal is to maximize its probability of survival, and who takes account of all relevant information about its environment.

7.6 Risk Preference

Our fifth parting-of-ways argument concerns decision-making under risk, that is, where an action's payoff depends on the state of the world, over which there is a known probability distribution. The standard definition of rational choice in this context is expected utility (EU) theory. On this approach, an agent has a utility function over the alternative outcomes, and evaluates gambles (lotteries) in accordance with their expected utility. Thus consider the lottery 'win \$10 with probability $\frac{3}{4}$; \$20 with probability $\frac{1}{4}$'. The agent's evaluation of this lottery is $\frac{3}{4} \cdot u(\$x + 10) + \frac{1}{4} \cdot u(\$x + 20)$, where $\$x$ is initial wealth and u is their utility function for money. Rationality then requires an agent to prefer one lottery to another if and only if the former has greater expected utility than the latter. Further detail on EU theory is given in chapter 8, section 8.2.

Though EU theory is a flexible framework, able to capture a diverse array of risk attitudes, it has descriptive shortcomings. A large body of research beginning with Allais (1953) has shown that peoples' risk preferences systematically violate EU maximization. Thus for example Kahneman and Tversky (1979) found that agents are often 'loss averse', or treat losses and gains asymmetrically, systematically overweight small probabilities, and have risk preferences that depend on the probabilities involved. (Their 'fourfold pattern' states that agents are risk-averse for high-probability gains or low-probability losses, but risk-seeking for low-probability gains and high-probability losses.) Similar violations have been found in non-human animals (Kagel *et al.* 1995). Given that risk preferences are likely to have evolved, a natural suggestion is that there may be an evolutionary basis for these rationality violations; and arguments to this effect have often been made.[16] If valid, such arguments would illustrate a striking parting-of-ways: rational and biologically optimal risk preferences would fail to coincide.

To understand how such a parting-of-ways is possible, recall the discussion of Bayesian rationality in chapter 6. Based on a simple model of optimal behaviour in an uncertain environment, we found that evolution would lead an organism to behave like an expected utility maximizer with utility = reproductive output. However, our model rested on a number of assumptions that, when relaxed, opens the door for a parting-of-ways argument. Such arguments are of two main sorts. The first sort relaxes the assumption that expected reproductive output is the right criterion of evolutionary success, or definition of fitness. The second sort retains this assumption, but tells a

[16] See for example Robson (1996), Brennan and Lo (2011), McDermott *et al.* (2008), Mallpress *et al.* (2015), and Hintze *et al.* (2015).

more complex story about how an organism's food or energy payoffs, over which its risk preferences are defined, translate into expected reproduction. I examine them briefly in turn; more detailed analysis is in the next chapter.

The long-term fate of a phenotype often depends not just on its expected reproductive success, but also on its variability in success. This point has long been recognized in general evolutionary biology, and, beginning with the seminal work of Robson (1996), has also been used to argue that violations of EU maximization can evolve. Crucial to such arguments is the distinction between 'idiosyncratic' and 'aggregate' risk. To understand this distinction, suppose that an organism performs an action that induces a lottery over possible levels of reproduction, for example, 'receive 1 offspring with probability $\frac{1}{2}$, 3 offspring with probability $\frac{1}{2}$'. Idiosyncratic risk means that each organism faces this lottery independently of others, that is, there is a separate flip of a fair coin for each. Aggregate risk is the opposite extreme: one coin is flipped for the whole cohort. In reality, organisms face both sorts of risk in different combinations. For example, the risk of a severe winter is aggregate, affecting many organisms at once, but the risk of getting struck by lightning is idiosyncratic—it will be independent across organisms, or nearly enough.[17]

To see why this distinction matters, consider the following widely used example.[18] A population contains two types, *A* and *B*, each with 1,000 members. Type is hardwired and perfectly inherited; generations do not overlap. Each *A* type definitely leaves two offspring per generation, while each *B* leaves either one or three offspring with equal probability. Suppose first that the *B*s face purely idiosyncratic risk. Then, neither type has an evolutionary advantage. In the second generation there will be $2,000A$ and almost exactly $2,000B$. Thus each generation, the *A*s will double and the *B*s will almost exactly double; so neither will predominate in the long-run. But matters are different if the *B*s face purely aggregate risk. In the second generation there will be $2,000A$ and either $1,000A$ or $3,000B$, with equi-probability. Even though both types have an expected reproductive output (or arithmetic mean) of two offspring per capita, over many generations the *A* type will predominate. This is because the geometric mean output of the *A* type exceeds that of *B*, owing to its lower variance.[19] Because reproduction is a multiplicative process, the type with the higher geometric mean has the long-run advantage, so will take over the population.

In this example, any *individual* organism of type *B* faces the very same lottery whether the risk is aggregate or idiosyncratic, so rationally should not discriminate between these cases if it cares about its expected number of offspring; however, evolution does discriminate. In principle, this allows risk preferences which violate

[17] The aggregate/idiosyncratic terminology comes from economic theory; in biology, an equivalent distinction is often drawn between environmental and demographic stochasticity.

[18] This type of example appears in Robson (1996), Sober (2001), and Bergstrom (2014), among others. For a rigorous treatment, see Robson and Samuleson (2010).

[19] The geometric mean of type *B* is $\sqrt{3}$ per capita, while that of *A* is 2. In general if two random variables have the same arithmetic mean, the one with the greater variance must have a smaller geometric mean.

EU theory to evolve. Thus Robson and Samuelson (2012) predict that evolution will lead agents to be more averse to aggregate risks than to equivalent idiosyncratic risks (thus potentially explaining why people are happy to drive on the motorway at night but are terrified of rare disease epidemics). Similarly, Brennan and Lo (2011) use the phenomenon of aggregate risk to show that agents who are loss averse, and who deliberately randomize their choices, may survive the evolutionary process. Such arguments raise a number of subtleties, analysed in chapter 8.

The second sort of parting-of-ways argument has a different logic. In behavioural ecology, researchers typically study organisms' preferences over gambles involving energy or food payoffs. The crucial issue is how such payoffs will affect the organism's reproductive success, or some component of it such as probability of survival. Even if the complication of aggregate risk is ignored, so expected reproductive success is the relevant criterion, the optimal risk preferences can be quite interesting. For example, a plausible assumption is that if an organism's energy reserves fall below a minimum threshold, it will die; but above a certain level, additional energy will not enhance fitness. Hence the optimal risk preference will depend on an organism's energetic reserves. At low reserves, the organism should be risk-prone—it pays to take a gamble even at bad odds if the alternative is certain death; but at high reserves the organism should be risk-averse. This is a standard prediction in the literature on risk-sensitive foraging in animals, and enjoys empirical support.[20]

This prediction may conceivably explain generic features of human risk preferences. But it does not necessarily take us beyond EU, contrary to what many authors imply, for the utility function may simply have convex and concave regions.[21] However, other authors have constructed explicit models, based on considerations of how expected reproductive success depends on energetic payoff, which do imply violations of EU maximization. A recent example is Mallpress *et al.* (2015). Their model combines the point that an organism's optimal risk preference will depend on its energy reserves with a second, less familiar point: the environment will often exhibit temporal auto-correlation, that is, its current state may predict the future state. They consider an organism foraging in a given environmental state, which yields a deterministic background rate of energy gain *d*. Each time period the organism can either accept or reject a gamble, whose expected payoff is equal to *d*. Remarkably, Mallpress *et al.* (2015) find that, in a wide range of parameter space, optimal risk preferences will satisfy the fourfold pattern of Kahneman and Tversky (1979).

To see why temporal auto-correlation matters, notice that it complicates the simple prediction that at low (high) reserves, an organism should always be risk-prone (averse). For an organism's expected lifetime reproductive success depends not just on its current energy reserves but also on future reserves; these in turn depend on how good the future environmental state is, which itself depends on whether the current

[20] See Bateson (2002) for a survey.
[21] Authors who imply this include Mishra (2014) and McDermott *et al.* (2008).

environmental state is expected to improve or deteriorate. Thus whether an organism should prefer a gamble, such as 'ten units of energy with probability $\frac{1}{2}$, 0 otherwise', to the certain receipt of five units, depends not just on its internal energy reserves, but also on the current environmental state and the degree of temporal auto-correlation. This complex dependence is what drives the unusual results of Mallpress *et al.* (2015).

Let us stand back and ask what these parting-of-ways arguments imply for the evolution–rationality connection. Regarding the first dimension, it appears that biological evolution can in principle lead to irrational risk preference (presuming we accept that EU is the rational ideal). Regarding the second dimension, it appears that the organism-as-rational-agent heuristic can potentially mislead. If we try to apply this heuristic in the context of risk, we would naturally conclude that an organism whose risk preferences have been optimized by natural selection will behave like an EU maximizer. But this is not necessarily true.

Two caveats are needed here. First, the second sort of parting-of-ways argument is less serious than the first, for the organism-as-rational-agent heuristic. For in the second sort of argument, for example, the Mallpress *et al.* (2015) model, the organism does still care about its expected (lifetime) reproductive success; it is just that this quantity bears a highly indirect relation to immediate energetic reward. So the heuristic fails if we treat the organism as trying to maximize its energetic payoff, and evaluating gambles over payoffs in accordance with EU. But instead, we can apply the heuristic by thinking of the organism as having the goal of maximizing expected lifetime reproduction, and having devised a rational strategy to that end; it is just that this strategy involves making choices over gambles that are non-rational when considered in isolation. So applied carefully, the organism-as-rational-agent heuristic is still valid. The dialectical situation here is similar to that of section 7.5, in relation to intransitive choice.

Second, with regard to the first sort of argument, it is natural to think that rationality and biological optimality can again be made to coincide by suitable choice of utility function. In the presence of aggregate risk, if utility is defined as the logarithm of offspring number, that is, if an organism values additional offspring in inverse proportion to how many it already has, then surely natural selection will lead to EU maximizing behaviour after all? (The point here is that maximizing the geometric mean of a random variable is equivalent to maximizing the expected value of its logarithm.) That is, perhaps this parting-of-ways argument can be reduced from the 'evolution $\Longrightarrow$ irrationality' type to the 'utility $\neq$ fitness' type? This suggestion is partly right, but as we shall see in chapter 8, it is not the whole story.

7.7 Inter-temporal Choice

Our sixth parting-of-ways argument concerns inter-temporal choice, in which an agent must decide how current rewards should be traded off against future rewards.

How such trade-offs are actually made, by humans and animals, is the subject of much empirical investigation. But what does *rational* choice amount to in an inter-temporal context? Some philosophers have argued that rationality requires us to value a reward identically irrespective of when it will be enjoyed.[22] However, this is a heroic move, which would diagnose virtually everyone as irrational, as empirically people invariably prefer an earlier over a later reward, that is, they discount the future. The traditional economist's view is that discounting is not irrational per se, as long as the agent discounts exponentially. This means that the agent's current evaluation (at time $t = 0$) of a reward of value V to be received at time t should equal Ve^{-rt}, where r is the agent's constant discount rate.

What is so special about exponential discounting? If follows from the requirement that an agent's preferences satisfy a condition called 'stationarity' (Fishburn and Rubinstein 1982). This means that an agent's current preference between receiving a reward at time t and and at a later time $t + \Delta t$ depends only on the interval Δt, and not on how far t is from the present. Thus if you prefer \$100 today to \$120 tomorrow, you should prefer \$100 in one year's time to \$120 in a year and a day. Intuitively this makes sense, since otherwise when a year has nearly elapsed, you will regret your previous choice, providing your (instantaneous) utility function for money is unchanged. Indeed the most powerful argument for exponential discounting is that any other way of discounting the future leads to 'time inconsistency': the agent will make choices that their later self will wish to repudiate (Strotz 1956).

Despite its rational credentials, an abundance of empirical evidence shows that humans do not in fact discount in exponential fashion. The usual finding is that the rate at which an agent discounts a future reward is not constant, but rather declines as the time to the reward increases. Thus in the previous example, many people will take the \$100 today over \$120 tomorrow, but nonetheless choose \$120 in a year and a day to \$100 in a year—perhaps feeling that an extra day will make no difference once they have already waited for a year. This phenomenon is known as 'preference reversal', and appears experimentally robust; it and other violations of the exponential model have motivated alternative descriptive models of inter-temporal choice. One such model is hyperbolic discounting, according to which future rewards are discounted by a factor of the form $\frac{1}{1+rt}$. On this model, with $r = 0.2$ and counting time in days, a dollar tomorrow is worth only 83.3 cents today; but a dollar in a year and a day is worth virtually the same as one in a year.

Violations of exponential discounting, including preference reversals, have been found in many animal experiments, with diverse species.[23] A standard experimental protocol involves a choice between a small reward after a shorter interval, and a longer

[22] For example, J. Rawls (1971) writes: 'rationality requires an impartial concern for all parts of our life. The mere difference of location in time, of something's being earlier or later, is not a rational ground for having more or less regard for it' (p. 293).

[23] See for example Stevens and Stephens (2010).

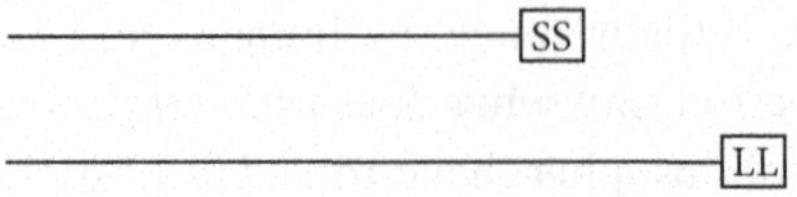

Figure 7.3. Choice between two rewards

reward after a longer interval (Figure 7.3). This tests the animal's self-control—can it overcome temptation and hold out for the larger reward?—and may bear some resemblance to real-life foraging decisions. One typical finding is that when the time to the small reward is very short, the animal is impulsive, choosing *SS*; but when a fixed delay is added to both intervals, a preference reversal occurs and the animal chooses *LL*. As with humans, it appears that animals' inter-temporal choices conform more closely to the hyperbolic than the exponential pattern.

Do the inter-temporal choices, or choice dispositions, of humans and animals have an evolutionary explanation? Many authors have suggested that they do. If so, then presuming that exponential discounting is accepted as the normative ideal, this would amount to a striking parting-of-ways: the rational and the biologically adaptive would fail to coincide. A full survey of the many models for the evolution of discounting will not be attempted; rather I focus on some key points.

If we ask why an organism discounts the future at all, the evolutionary rationale seems clear. Sooner is better than later, since there is a chance that a later reward may not materialize, for example, a foraging bout may be interrupted, or the organism may die. Similarly, reproducing today is better than tomorrow, other things being equal, as tomorrow may never come. But can evolutionary considerations explain actual patterns of inter-temporal choice, such as the fact that discount rates seem to decrease as the time to the reward increases? One long-standing idea in behavioural ecology is that there is a simple explanation: foraging animals have evolved to maximize their rate of energetic intake per unit time. Thus faced with the choice in Figure 7.3, between *SS* after an interval of t or *LL* after t', the animal will prefer *SS* if and only if:

$$\frac{SS}{t} > \frac{LL}{t'}$$

The significance of this is that when a fixed interval Δt is added to both t and t', the inequality may reverse (and must do eventually, if Δt is made big enough):

$$\frac{LL}{t + \Delta t} > \frac{SS}{t' + \Delta t}$$

Thus maximization of rate of energetic intake—a widely used proxy for fitness in foraging studies—can explain why preference reversal will occur, and more generally can explain why an organism's discount rate should decline with time, as has been found empirically. This observation prompted Hagen *et al.* (2012) to argue that what is irrational in economic terms may very well be biologically adaptive.

This point may be part of the story, but many alternative explanations of inter-temporal discounting have been proposed. For example, Stephens (2002) argues that constraints on discrimination may play a role. An organism can more easily distinguish between smaller time intervals, for example, five versus eight seconds, then larger ones, for example, twenty-five versus twenty-eight seconds, as the former difference is more salient. Thus greater error will affect long-term comparisons, so impulsive choice rules (which lead to strongly present-biased choices and declining discount rates) will be advantageous. The logic here is familiar: when there are bounds on agent's cognitive/perceptual powers, apparently irrational choices can make sense.

A different class of arguments focuses on the hazard rate, that is, the chance that a delayed reward will not materialize, for example, because it decays, or is taken by a competitor. In the model of Fawcett *et al.* (2012), there is a fixed chance per unit time α that a future reward will not materialize, to which the animal is perfectly adapted. The animal's goal is to maximize its expected reward at the end of a fixed time period, during which it faces a repeated choice between *SS* and *LL* options. This would seem to favour exponential discounting: the present expected value of a reward V that is delayed by t units of time is $Ve^{-\alpha t}$, so surely the well-adapted animal should discount exponentially at the hazard rate α? However, the authors show that this is only true when there is no opportunity cost of waiting for the delayed reward, which is unrealistic. It is more likely that an animal that holds out for an *LL* reward will forgo other foraging opportunities as it waits. When this is factored in, it turns out that exponential discounting is no longer optimal, and preference reversals can occur.

A still different argument, also focusing on hazard rate, is given by Sozou (1998). He assumes that an organism (or agent) can learn about the hazard rate within its lifetime, rather than being adapted to it by evolution. Thus as an organism waits for a delayed reward, as time goes by and no hazard occurs, it updates its beliefs about the hazard rate in a Bayesian manner. Sozou shows that depending on what an organism's prior probability is that a given reward will still be available at a later time t, its posterior probability may itself have a hyperbolic form. To see how this can lead to preference reversals, consider again the choice between the *SS* and *LL* options. Believing the hazard rate to be substantial, the agent prefers *SS*, for though it yields a smaller reward it is much more likely to materialize. But if a fixed time interval Δt is added to both options, the larger *LL* option is preferred. For both rewards are further in the future, so although both are less likely to materialize, the relative likelihood of *LL* materializing, vis-à-vis *SS*, is higher. In short, uncertainty about the hazard rate, plus Bayesian updating from a suitable prior probability, imply that hyperbolic time-preference is optimal, in this model.

What do these arguments show? Taken at face value, they suggest that irrational choices can sometimes be adaptive. However, this interpretation could be questioned, for two reasons. First, some of the explanations make the choices in question seem only apparently irrational. This is because they invoke additional factors (opportunity costs, repeated choices, uncertain hazard rates) that are not present in the traditional human experiments on time-preference, for which exponential discounting is supposed to be the normative ideal. This is not a criticism—such factors may well have been relevant in the environments in which decision-making, human and non-human, evolved. The point is rather that it defuses our sense that the behaviours that evolve in these models are normatively inappropriate. Organisms evolve time-preferences that coincide with those that a rational human would have, *if* they faced the environmental and informational constraints posited by the model and their goal was to maximize the relevant fitness proxy. This raises a more general issue about how parting-of-ways arguments should be interpreted, discussed in the next section.

Second, whether exponential discounting is a true rationality requirement for human agents is in any case a matter of debate. Despite the threat of dynamic inconsistency, some authors argue that people are not rationally required to discount exponentially. A salient point here is that when experimental human subjects have it explained to them that their choices are time-inconsistent, they do not regard themselves as having made a mistake (unlike with violations of transitivity, for instance) (Frederick *et al.* 2002). Also relevant is the point that people often 'pre-commit' to a plan of action, precisely to prevent their future self from exercising a choice that from today's perspective is sub-optimal (Ainslee 1992). Such people seem to know that they are not exponential discounters, and respond rationally to this knowledge. Thus compared with some of the other rationality norms we have discussed (avoidance of dominated strategies, transitivity, EU maximization), the status of exponential discounting is less clearcut.

To sum up: many models suggest that inter-temporal choices that deviate from exponential discounting may be biologically advantageous. But whether this represents a clear parting-of-ways is open to question, both because the status of the norm is itself moot, and because some of the models invoke factors that, if known to be present, would cause a rational human to deviate from exponential discounting too.

7.8 Upshot

We have examined six parting-of-ways arguments, or really families of arguments, from the literature, all purporting to show that adaptive behaviour can sometimes involve violation of (alleged) rationality norms. In each case, we have asked whether the mismatch between adaptiveness and rationality is real or apparent, explored possible ways of eliminating it, and considered the implications for both dimensions of the evolution–rationality connection. The results are summarized in Table 7.4. In one case—risk—our conclusions are provisional, pending further analysis in the next chapter.

Do any general morals emerge from these arguments? Or do they merely show that with sufficient ingenuity, modellers can find a biological advantage for almost anything? I think we can say a bit more than this. One moral is that if we define

Table 7.4. Parting-of-ways arguments

Phenomenon	Norm violated	Possible resolutions
Cooperation	Avoid strictly dominated strategies	*Re-define, adopt*
Fairness	Avoid weakly dominated strategies	*Re-define*
Trust	Maximize objective payoff	*Re-define*
Intransitivity	Avoid intransitive choices	*Change*
Risky choice	Maximize expected utility	*Re-define, adopt* (?)
Inter-temporal choice	Discount exponentially	*Reject*

Key: *re-define* = re-define utility function; *change* = change state-space; *adopt* = adopt different fitness measure; *reject* = reject norm.

rational behaviour sufficiently precisely, as consistency with specific rational choice norms, then it is an open theoretical question whether natural selection will favour rationality, not a given. This may seem obvious; however, it is a useful corrective to the popular assumption that being rational is obviously biologically advantageous. (Recall Dennett's statement that 'natural selection guarantees that most of an organism's beliefs will be true, its strategies rational' (1987, p. 7).) This assumption is *prima facie* appealing, but closer examination shows that it does not always hold.

A second moral is that biological explanations of observed departures from ideal rationality in humans, practical and theoretical, do not necessarily need to invoke cognitive or computational limitations, or constraints on the power of adaptation, such as a mismatch between ancestral and modern environments. Such limitations and constraints may be relevant; however, to assume that they must be at work is to assume that ideal rationality is the biological optimum, so deviations from it must reflect the inability of agents to compute the optimal course of action, or else failures of perfect adaptation. However, the parting-of-ways phenomenon shows that this assumption is not universally valid, since in some cases there can be an adaptive advantage to irrationality, for example, playing weakly dominated strategies or violating time-consistency.

A third moral is that the simplest way of linking rationality and biological adaptiveness, namely equating an agent's utility for an action with the additional number of offspring that the action brings, will not always work. But it is often possible to salvage the link by choosing a more complicated utility function, or by changing the state-space of the decision problem, or by adopting a different fitness measure, as we have seen. However, there is no guarantee that this can be done in every case, and the appropriate re-formulation will typically depend on model assumptions.

A fourth moral concerns the organism-as-agent heuristic—the anthropomorphic assimilation of a well-adapted organism to a rational agent consciously pursuing a goal. This heuristic can be a powerful way of understanding evolved behaviour, as we saw in our discussion of agential thinking in chapter 1. It is tempting to suppose that the heuristic *must* be applicable, as long as the organism's behaviour is adaptive, and its goal is suitably specified. But when 'rational' is independently defined, and invested with a sufficiently precise meaning, this is not necessarily so. The link between what is adaptive and what is rational can break down, and cannot always be restored by suitable choice of goal, that is, utility function. Thus the heuristic must be treated with care, and not regarded as a definitional truth.

Finally, I want briefly to consider an issue that has reared its head already. If a successful parting-of-ways argument can be given, to the effect that adaptive behaviour may violate a particular rationality criterion, should this lead us to question the normative status of that criterion? For example, if there is a good biological reason why organisms have evolved risk preferences that violate the expected utility axioms, or choice dispositions that violate transitivity, does this undermine the idea that conformity to these norms is a demand of rationality?

Not necessarily. As discussed in section 6.7, there are powerful non-evolutionary arguments for why a rational agent should obey the usual norms of E- and P-rationality, irrespective of whether doing so is biologically adaptive. So the fact that violations of these norms can sometimes be adaptive does not in itself threaten their status as norms. However, some parting-of-ways arguments trade on the point that an organism's decision-making mechanisms may be well-suited for the environment in which they evolved, but yield odd results in artificial experimental set-ups, such as ones that involve single-shot choices. (This is a key part of the 'ecological rationality' concept.) In such cases, the organism may seem exonerated from the charge of irrationality—surely the fault lies with the experimenter, not the organism?

Two points should be noted here. First, it makes a difference whether the organism in question is human or not. Certainly it is odd to treat a non-human organism, which lacks a conceptual understanding of the choice task it has been presented with, as having done something 'wrong' if it is induced to make a mal-adaptive choice by being placed in a non-naturalistic experimental setting. But for a human who does understand the task, matters are different. Even if humans are psychologically disposed, as a result of biological evolution, to violate norms of rational choice and belief in experimental tasks, this does not make it right to do so. To insist otherwise is to deny the possibility of an a priori justification for those norms which is independent of their adaptive value (or lack thereof). This general point holds good even if we do not agree on exactly what the norms are.

Second, the ecological rationality defence is in danger of equivocating on the notion of reason. There may be a good biological reason, or 'evolutionary rationale', for why an organism exhibits the behavioural choices that it does, but that does not necessarily make the choices themselves reasonable or rational. The conceptual link between what is adaptive and what is rational, and the formal link between maximization of fitness and of utility, does not mean that one may be reduced to the other. This point is fairly obvious, but is obscured by the tendency in some quarters to treat 'rational' and 'adaptive' as synonymous in a biological context. This tendency should be resisted, not least because it threatens to pre-judge the issue of whether the parting-of-ways phenomenon is possible at all.

7.9 Conclusion

Given the abstract parallel between fitness and utility, it is natural to think that rational and biologically optimal behaviour will coincide, as many authors have assumed. However, the models surveyed here suggests that matters are not always so simple: when rationality is defined sufficiently precisely, it can in principle part ways from biological adaptiveness. How robust this parting-of-ways phenomenon is is a tricky question, since the models have been deliberately designed to produce it. Moreover, the parting can sometimes be eliminated by suitable choice of utility function/fitness measure, or by re-framing the decision problem; and in some cases the rationality norms may themselves be queried. The overall moral is that it is not a given that natural selection will favour rational over irrational behaviour. So the organism-as-rational-agent heuristic must be treated with care, not regarded as a definitional truth, even on the assumption that the organism's behaviour has been optimized by natural selection.

8

Risk, Rational Choice, and Evolution

8.1 Introduction

In previous chapters, we noted that ideas from rational choice theory often have analogues in evolutionary biology, due to both cross-fertilization and convergence. One example is the problem of optimal choice in the face of risk, that is, where the outcome of an action is probabilistic rather than certain. How a rational agent should behave in such a context is a classic problem in decision theory. This problem finds a parallel in two areas of biological discussion. The first is risk-sensitivity in behavioural ecology, which studies animals' foraging behaviour when resources are variable. The second is evolution in variable environments, where the payoff to a genotype or phenotype depends on the environmental state, over which there is a given probability distribution. There the question concerns what is optimal in the biological sense, rather than the rational sense.

This chapter explores the connections, thematic and formal, between the decision-theoretic and the biological discussions of risk. Remarkably, it turns out that a number of conceptual issues in traditional decision theory, some of which are of philosophical interest in their own right, have biological parallels. These include the meaning of risk-aversion, the use of decision weights other than the true probabilities, and the tactic of re-formulating the state-space to eliminate apparent anomalies of choice.

The chapter has three aims. The first is to explore these parallels, and to reflect on what if anything they teach us. This ties in with a number of themes from previous chapters, in particular the evolution–rationality connection and the fitness-utility relation from chapter 6. The second aim is to use risk as a test-case for assessing our previous ideas about agential thinking in biology, developed in chapter 1. The third aim is to elaborate further on the evolution of risk preferences, a topic briefly treated in section 7.6 in relation to the parting-of-ways concept.

The structure of this chapter is as follows. Section 8.2 rehearses expected utility (EU) theory, the standard normative theory of decision under risk, and considers some objections to it. Section 8.3 discusses risk-sensitive foraging theory, emphasizing its structural similarity to EU theory. Section 8.4 examines evolution in variable environments, and asks whether evolution's 'choices' correspond to those of a rational agent. Section 8.5 asks whether organisms with irrational risk preferences could ever have an advantage, outlines a biological version of the Allais paradox, and considers the implications for agential thinking. Section 8.6 examines bet-hedging and mixed strategies, focusing on the vexed issue of whether such strategies imply individual irrationality or not. Section 8.7 concludes.

8.2 Expected Utility and its Discontents

The origins of EU theory lie in D. Bernoulli's observation that in games of chance, agents' choices do not maximize expected monetary value (Bernoulli 1738). This observation, made dramatic by his 'St. Petersburg paradox', today enjoys the status of a well-confirmed fact. For example, agents typically choose \$5 for sure over a gamble which pays either \$10 or nothing depending on the flip of a fair coin, despite both having the same expected monetary value; that is, they have risk-averse preferences. Bernoulli argued that agents instead maximize their expected utility, where 'utility' is a measure of psychological satisfaction that scales non-linearly with wealth. If an agent's utility function for wealth is concave, maximization of expected utility will lead them to be risk-averse. To see this, note that if u is concave, then $u(\$5) > \frac{1}{2}[u(\$10) + u(\$0)]$ (Figure 8.1). Bernoulli suggested one particular concave function, namely log(wealth), which means that an agent values an additional unit of wealth in inverse proportion to how much they already have.

Bernoulli's theory was clever but raised certain questions. Why should utility be assumed quantifiable on a cardinal scale (as his theory requires)? Why should the utility function be logarithmic? And why should an agent care only about the expected utility of a gamble, rather than also caring about the variance in utility, for

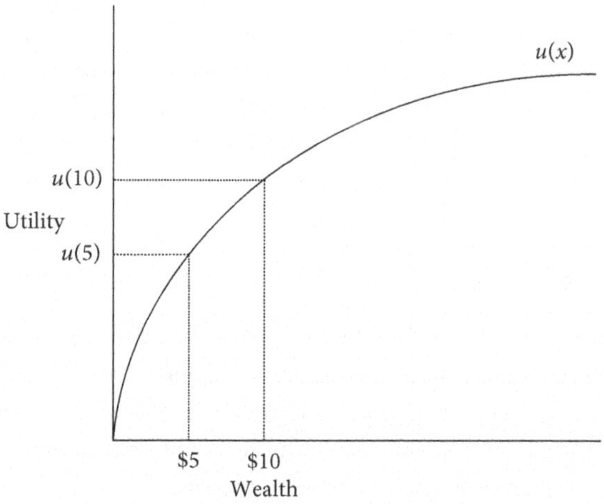

Figure 8.1. Concave utility function

example? Modern decision theory took the idea of EU maximization from Bernoulli but developed it in a different way, thus supplying answers, of a sort, to the first and third of these questions. The key idea is to take as primitive an agent's preferences, which are assumed to satisfy certain axioms, and to deduce the principle of EU maximization from that basis.

The canonical formulation of this idea is due to von Neumann and Morgenstern (1944). In their treatment, an agent is faced with a choice between lotteries, where a lottery is an objective probability distribution over a set of outcomes. These could be a set of monetary prizes, for example, whole dollar amounts from one to ten. A lottery specifies the probability that each outcome has of occurring; thus $(\$1, p_1; \$2, p_2; \ldots; \$10, p_{10})$ is the lottery in which amount $\$i$ is received with probability p_i. They then prove that as long as the agent's preferences satisfy a small number of fairly compelling axioms, the agent behaves as if they have a cardinal utility function on the outcomes and choose between lotteries in accordance with maximum expected utility. The key axiom is the independence axiom, which says that if an agent prefers L_1 to L_2, then they should prefer the lottery 'L_1 with probability p, L_3 otherwise' to 'L_2 with probability p, L_3 otherwise'; that is, the agent's preference for one lottery over another should be unchanged if each is mixed in identical proportions with a third lottery.

In EU theory, an agent's attitude towards (pecuniary) risk is reflected in the shape of their utility function u for money. If u is concave (i.e. $u'' < 0$) then the agent is risk-averse; if u is convex (i.e. $u'' > 0$) the agent is risk-seeking; while if u is linear (i.e. $u'' = 0$) then the agent is risk-neutral. This follows immediately if we define risk-aversion in the standard behaviourist way, namely strictly preferring $\$x$ for sure to any risky gamble with expected value of $\$x$, for any amount x.

Bernoulli's theory was unable to explain why an agent should care only about expected utility, rather than also attending to the variance in utility. On the von Neumann and Morgenstern theory, this question simply does not arise. That a rational agent should care only about the expected utility of a lottery, where rational is defined as having preferences that conform to the axioms, follows immediately from their representation theorem. Essentially, this is because utility is defined as the quantity that falls out of the theorem, rather than having an intrinsic psychological meaning.

8.2.1 Objections to EU theory

Since its inception, EU theory has been subject to criticism. One of the earliest critics, Allais (1953), argued that the independence axiom was neither normatively compelling nor empirically plausible, and that the equation of risk attitude with curvature of the utility function was conceptually problematic. These two points, both of which are still debated today, merit brief comment.

Empirically, there is evidence that people's preferences systematically violate the independence axiom. Allais originally illustrated this using a famous pair of examples.

Table 8.1. A version of the Allais paradox

	$\frac{2}{10}$	$\frac{1}{10}$	$\frac{7}{10}$
a	$1 million	$1 million	$1 million
b	$0	$5 million	$1 million
c	$1 million	$1 million	$0
d	$0	$5 million	$0

A variant on his first example goes as follows.[1] An agent is offered a choice between option *a*, which pays $1 million for sure, and *b*, which yields a 70% chance of $1 million, a 10% chance of $5 million, and a 20% chance of nothing. Many agents choose *a* in this sort of scenario, preferring the certainty of $1 million to the riskier option *b*, despite the latter's higher expected value. Next, the agent is offered a choice between option *c*, which yields a 30% chance of $1 million and a 70% chance of nothing; and *d*, which yields a 10% chance of $5 million and a 90% chance of nothing. Many agents choose *d* in this sort of scenario, perhaps reasoning that since they are unlikely to win anything anyway, they might as well take the chance of getting the $5 million. However, the independence axiom requires that *a* is preferred to *b* if and only if *c* is preferred to *d*. This is clear from the presentation of the options in Table 8.1. The $\{a, b\}$ and $\{c, d\}$ pairs are identical apart from in the third column, which yields the same outcome for both members of each pair, so should be irrelevant. It is easily verified that if an agent has these 'Allais preferences', they cannot be represented as maximizing EU. Such preferences have been found in both humans and animals.[2]

Further violations of EU theory have also been discovered, as discussed briefly in section 7.6. These include the certainty effect, in which agents overvalue probabilities that are near to maximal, and the common-ratio effect, in which an agent's preference between lotteries changes when the probabilities of non-zero outcomes are multiplied by a common factor—for example, preferring a 10% chance of $\$x$ to a 20% chance of $\$y$, but a 2% chance of $\$y$ to a 1% chance of $\$x$. Another violation is having a strict preference for randomization, which occurs when an agent prefers a gamble between x and y to the certain receipt of x, and also to the certain receipt of y; see section 8.6. Such phenomena cast doubt on the descriptive validity of EU theory.

One response to this situation, popular among those who regard EU as the normative ideal, is to invoke a re-description strategy. In the Allais paradox, for example, it is sometimes suggested that if the outcomes are allowed to include feeling of regret, as well as specifications of pecuniary reward, then the anomaly disappears.[3] Thus

[1] I have used the actual amounts in Allais' original example but altered the probabilities. This leaves the logic unchanged, and facilitates the point made in section 8.5.1, on the biological Allais paradox.

[2] See Huck and Müller (2012) for data on humans, and Kagel *et al.* (1995) for data on animals.

[3] This move is suggested by Tversky (1975), Broome (1991), and Weirich (1986).

Table 8.1 shows lotteries *b* and *d* as sharing a common outcome of \$0 in the left-most column. However, in *b*, the outcome could be described as 'receiving \$0 when the certainty of \$1 million was available'; but this re-description obviously does not apply to the \$0 in *d*, as no certain reward is available from the $\{c, d\}$ pair. Therefore, it is argued, lotteries *b* and *d* do not share a common outcome in the left-most column, so the independence axiom is not violated after all.

In this case the re-description strategy has a certain plausibility—as feelings of regret may well be part of the explanation of why people have Allais preferences. But without a limit on the permissible re-descriptions, the strategy threatens to render EU theory vacuous; we discussed this issue in relation to transitivity in section 7.5. In section 8.6.2 we will find that a biological analogue of the re-description strategy arises in relation to biologically optimal risk preferences.

The second objection to EU theory is conceptual rather than empirical. Beginning with Allais (1953), the suggestion has often been voiced that the orthodox EU treatment of risk is flawed, since it conflates attitude towards wealth with intrinsic risk attitude. Thus consider again an agent who prefers \$5 for sure to a fair bet on \$10 or nothing. Intuitively such a preference might reflect the fact that the agent's utility function for wealth is concave, that is, increasing units of wealth bring diminishing marginal utility; but it might not. Even if their utility function is linear, they might prefer the certainty of \$5 because they have an intrinsic dislike of risk. But EU theory appears to countenance only the first explanation, as it identifies being risk-averse with having a concave utility function.

One version of this objection involves rejecting the 'representationalist' view of utility in favour of the older idea that it is a real psychological quantity, in principle measurable on a cardinal scale even in contexts not involving risk. This was Allais' own view. He argued that a rational agent may take account of the entire distribution of cardinal utilities associated with a risky gamble, not just the expected utility. In particular, the variance of the utilities will matter to the agent, Allais argued. A risk-averse agent will wish to minimize their variance in utility, so will trade off the variance against the expected utility in their evaluation of a gamble.[4]

Orthodox EU theorists regard this argument as confused, stemming from a mistaken assimilation of utility in the sense of von Neumann and Morgenstern to the sort of psychological utility with which Bernoulli was concerned. On their view, an agent's (vNM) utility function is not meant to explain their preferences but merely to represent them; so concave utility is not a hypothesis that might explain why an agent exhibits risk-averse preferences, to be contrasted with other hypotheses. No meaning attaches to the idea that an agent might care about the variance in utilities of a gamble, on the orthodox view.

[4] Allais did not say how this trade-off should be done, a point that his critics were quick to pick up on. See the papers in Allais and Hagen (1979).

This debate raises complex philosophical issues.[5] But whatever one's view on whether utility has an intrinsic psychological meaning, the intuition that attitude towards wealth is one thing, attitude towards risk another, is a powerful one. Moreover, this intuition is part of the motivation behind 'non-expected utility theory'. This is the generic name for a class of decision models that abandon the independence axiom, of which prospect theory is perhaps the best known.[6] The main aim of these models is to produce a descriptive theory that accommodates empirical violations of EU theory such as the Allais preferences. But a secondary motivation is to separate risk attitude from curvature of the utility function.

To see how non-EU theories handle risk attitude, consider a lottery of the form $(\$1, p_1; \$2, p_2; \ldots, \$10; p_{10})$. In EU theory an agent evaluates this lottery by $\sum p_i \cdot u(\$(x + i))$, where x is initial wealth. In many though not all non-EU models, the corresponding evaluation criterion can be expressed in the form $\sum w_i \cdot u(\$(x+i))$, where w_i is the *decision weight* of outcome i, with $\sum w_i = 1$. That is, agents maximize the weighted average of the utilities of the outcomes, using weights that need not equal the true probabilities. The decision weight of an outcome typically depends partly on its probability and partly on other factors, for example, the probability of getting another outcome at least as good. Certain non-EU models, such as prospect theory, introduce a further twist by taking the utility function u to be defined on gains or losses, rather than total wealth.

In non-EU theory a distinction between risk attitude and shape of the utility function arises immediately. This is easy to see when the evaluation criterion has the generic form of the previous paragraph. The shape of u does not fully determine whether an agent will exhibit risk-averse preferences; this also depends on the weighting function w. Thus there are two distinct ways that risk-averse preferences can arise. One is concave utility; the other is 'probabilistic risk-aversion', which arises from using decision weights that diverge from the true probabilities (Wakker 1994). Thus the intuitive idea that attitude towards risk is one thing, attitude towards wealth another, is preserved.

8.3 Risk in Evolution

The topic of risk arises in evolutionary discussions in two ways. First, a species' environment may vary stochastically, so a trait's effect on survival and reproduction may itself be stochastic. This general point applies to traits of all sorts, not just behaviours. For example, growing thick fur may be advantageous for a mammal in a cold winter but not otherwise. So natural selection must 'choose' the optimal

5 I discuss some of these issues in Okasha (2016). See Buchak (2013) and Dietrich and List (2016) for other perspectives on the issue.

6 Early work in non-EU theory was by Kahneman and Tversky (1979), Machina (1982), and Quiggin (1982). For reviews, see Starmer (2000) and Wakker (2010).

level of fur thickness in the light of the probability of a cold winter and the fitness consequences of the different options. Second, many organisms make behavioural choices within their lifetime. Thus a butterfly must choose whether to lay its eggs on old or young leaves, and a bird must choose whether to forage more intensively in the morning or afternoon. Such choices often involve risk, since their fitness consequences are probabilistic, not certain. So to be adaptive, an organism's behaviour will need to take account of this risk.

Both of these issues may be conceptualized in decision-theoretic terms. In the former, the decision-maker is natural selection and the (metaphorical) choice is between alternative traits; in the latter, the decision-maker is an individual organism and the (literal) choice is between alternative actions. (This corresponds to our previous distinction between agential thinking of type 2 and of type 1.) In both cases, we can ask whether the choices in question will conform to the principles of rationality, as enshrined in EU theory, and if not why not. We can also consider the two issues at once, by thinking of an organism's evolved preferences as having themselves been 'chosen' by natural selection. Let us begin with the second issue.

8.3.1 Risk-sensitive foraging

To fix ideas, consider a typical example from the literature on risk-sensitive foraging. An organism has two foraging options. Option X is safe—it guarantees five units of food in a given time period. Option Y is risky—it brings either nine units of food or one unit, with equal probability. For example, the organism could be an insect and X and Y flower types with different nectar contents. This is conceptually similar to a standard rational choice problem, with food (or energy) replacing money as the reward currency. Which should the organism choose? Both options have the same expected reward, namely five units of food per time period, so one might think that they are equally good. However, experiments show that when presented with choices of this sort, organisms exhibit a definite preference—they usually choose the safe option. Such risk-averse preferences have been found in birds, mammals, and insects.[7]

Why might it be advantageous for an organism to be risk-averse? The most straightforward reason is that an organism's biological fitness may scale concavely with energy intake. That is, additional units of food translate into increased reproductive success with diminishing returns (Figure 8.2). Empirically this is fairly plausible, and is a standard assumption in models of risk-sensitive foraging.[8] It implies that when faced with a choice between two options with the same expected reward, such as X and Y in the foregoing example, the organism should choose the option with the lower variance in reward in order to maximize its expected fitness, that is, X. In principle, there is thus a simple adaptive explanation for why organisms are often risk-averse when faced with variability in resource acquisition.

[7] See Bateson (2002) for a review of the empirical findings.
[8] See for example Stephens (1981) or Real (1992).

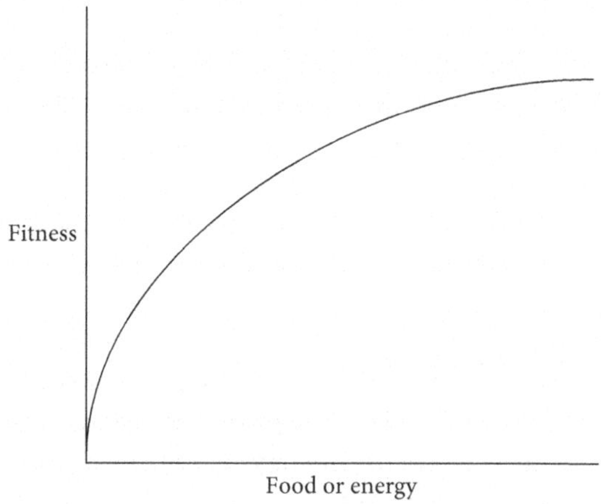

Figure 8.2. Concave fitness function

The logic of this explanation is pleasingly isomorphic to how traditional EU theory accounts for humans' risk preferences over monetary gambles, that is, positing a non-linear relationship between money and utility. As we have seen, this was part of Bernoulli's original motivation for the idea of EU maximization. If we replace food with money, and biological fitness with utility, the two explanations map onto each other exactly. In both cases, the non-linearity means that to maximize expected utility/fitness, a rational agent/evolved organism should be sensitive to variance in reward as well as expected reward, as empirically they are. The adequacy of both explanations to their respective data can and has been questioned, but for the moment it is the isomorphism that I want to stress. It provides a clear example of the organism-as-rational-agent idea, and the utility = fitness hypothesis, discussed previously, actually at work in biology. Indeed, proponents of risk-sensitive foraging theory have sometimes explicitly drawn on EU theory.[9]

This abstract parallel between EU theory and risk-sensitive foraging, while interesting, should not blind us to the differences. The former is part of a proximate explanation (or description) of behaviour, while the latter is concerned with adaptive, or ultimate, explanation. Moreover, utility is a notional quantity while biological fitness is a real one, as stressed in section 6.6. Indicative of this is that the shape of an agent's utility function is a primitive in EU theory, reflecting the subjective tastes of the agent; but models of risk-sensitive foraging often try to explain why the plot of fitness against food reward is non-linear, rather than just positing it.

Despite these points, and despite the empirical challenges which both theories face, I think that the parallel is instructive, particularly in regard to the treatment of risk-aversion. It provides a concrete illustration of why the organism-as-rational-agent

[9] See in particular Real (1992) and Real and Caraco (1986).

heuristic is such a compelling way to think about evolved behaviour. Faced with variability in resource acquisition, an organism whose choices conform to the predictions of risk-sensitive foraging theory behaves in exactly the way that a rational agent, as defined by EU theory, chooses between lotteries. Thus we can sensibly treat the organism as trying to achieve a goal and responding rationally to the uncertainty it faces. In this case the heuristic does genuine intellectual work, suggesting a conceptual framework that has yielded testable empirical predictions, so is not mere anthropomorphism.

It is tempting to go further than this and to see in risk-sensitive foraging a biological basis for the human risk preferences that are the subject matter of EU theory (and its non-EU modifications). For example, perhaps the ubiquity of risk-aversion towards monetary gambles is an evolved psychological trait which has its origin in the concavity of the function mapping food reward to reproductive fitness? If so, then the conceptual link between risk-sensitive foraging and EU theory would be no accident, but would reflect the fact that the former provides an evolutionary explanation for the patterns of preference that the latter tries to codify. However, this is an empirical hypothesis which may or may not be true.

8.3.2 A complication

The apparent isomorphism between the biological and the decision-theoretic analyses is complicated by one factor. Consider again the choice between option X (five units of food for sure) and Y (nine or one units with equal probability). Let us suppose that biological fitness is linear in food reward, so both X and Y yield identical expected fitness. This may seem to imply that an organism, to be adaptive, should be indifferent between X and Y. However, this need not be true. For there is a second reason why it may be advantageous for an organism to prefer X to Y, which is independent of the curvature of the plot of fitness against food reward.

The reason is this. In many circumstances, traits that lead to a high variance in fitness are penalized by natural selection. In particular, if the reproductive success of a trait varies stochastically, then expected reproductive success is not the sole determinant of the trait's evolutionary success—the variance matters too. (By the reproductive success of a trait, or genotype, we mean the average reproductive success of organisms with that trait, or genotype.[10]) This means that natural selection may favour an organism who prefers option X to Y, in our example, even if reproductive success is linear in food reward.

To see this point, suppose that an organism's preference between options X and Y is genetically hard-wired, and transmitted to its offspring. Since we are supposing that fitness is linear in food, we are effectively dealing with gambles over offspring. Thus

[10] Here it is crucial not to confuse the average reproductive success of a type, which involves averaging over individuals, with the expected reproductive success (of either an individual or a type), which involves taking expectation over possibilities.

choosing option X brings an organism 5 offspring per time-period; while choosing Y gets 9 or 1 offspring with equal probability. Let us equate a time-period with a generation. Further, let us suppose that the risk faced by organisms choosing option Y is aggregate; this means that in each generation, a *single* coin flip determines whether each leaves 9 or each leaves 1. The contrast is with idiosyncratic risk, in which each organism choosing Y faces an *independent* coin flip between 1 and 9 offspring. Recall the discussion of aggregate versus idiosyncratic risk from section 7.6.

That the risk in this example is aggregate is perfectly plausible, biologically. Suppose that the organism in question is an insect, and option Y is a flower type whose nectar content is either high or low in a given generation, with equal probability, depending on seasonal fluctuations in temperature. This means that in any generation, every organism who chooses Y will get 9 offspring if the nectar content is high that season, or 1 if it is low. In general, temporal variation in environmental conditions, such as fluctuations in the weather from year to year, gives rise to aggregate risk, while spatial variation gives rise to idiosyncratic risk. In practice both sorts of variation are common, so organisms typically face risks of both sorts in varying proportions.

The key point to note is that aggregate risk gives rise to between-generation variation in the reproductive success of a type, while idiosyncratic risk does not. (Since the preference for option X over Y is hard-wired, we can think of X and Y as competing types.) To see this, note that if each Y type faces an independent coin flip between 1 and 9 offspring, then as long as the population is large, the average reproductive output of the Ys in any generation will be very close to 5, by the law of large numbers. That is, idiosyncratic risk leads to variability in the reproductive success of individuals of the same type, but this averages out in a large population, while aggregate risk leads to variability in the (average) reproductive success of the type. The latter but not the former is penalized by natural selection.[11]

The reason why natural selection penalizes variability in a trait's reproductive success is explained in section 8.4. For the moment, the point is just this. When there is aggregate risk, as for example when the environment fluctuates stochastically over time, natural selection itself acts in an intrinsically 'risk-averse' way, in the sense that it favours traits which minimize the variance in (trait) reproductive fitness, for a given expected fitness. Thus variability in fitness is itself selectively penalized.

When the phenomenon of aggregate risk, as applied to organisms' behavioural choices, is taken in conjunction with the concept of risk-sensitive foraging, described previously, we see that there are two separate reasons why evolution may lead organisms to exhibit risk-averse behaviours. The first is a concave relation between a resource such as food and biological fitness; as we have seen, this implies that to maximize its expected individual fitness, an organism should choose a safe over a risky alternative with the same mean reward. The second is the fact that in stochastic

[11] This assumes a large population. In a finite population, the former sort of variability is also penalized. See Gillespie (1974, 1977) or Frank and Slatkin (1990) for discussion.

environments, traits including behaviours that have a high variance in (trait) fitness are selected against. Thus even if an individual's reproductive fitness is linear in food resources, risk-averse preferences may still evolve; for expected fitness is not the sole determinant of the outcome of evolution.[12]

There is an interesting parallel here with rational choice. In section 8.2 we saw that intuitively, there are two reasons why a rational agent might have risk-averse preferences for monetary gambles—a concave utility function for money and an intrinsic distaste for risk—which EU theory is often accused of conflating. These are analogous to the two reasons why evolution may lead organisms to have risk-averse preferences. In both cases, the first reason is a concave relationship between a resource (money or food) and some measure of value (utility or reproductive fitness); the second is the fact that the dispersion of the distribution of those values, as well as the expected value, sometimes matters. It is striking that this distinction between concave utility and intrinsic risk-aversion, which cannot be made within EU theory, finds an apparent parallel in biology.

This parallel is suggestive, but I do not want to make too much of it. However, it prompts one important question. When we take into account environmental variability, and the aggregate risk to which it can give rise, what are the consequences for the utility/fitness parallel and for agential thinking? Given that expected fitness is not always the criterion of evolutionary success, does this mean that evolution may favour organisms whose choice behaviour violates EU maximization? I return to this in section 8.5, via a detour.

8.4 'Mother Nature' and Geometric Mean Fitness

In section 8.3 we discussed the evolutionary consequences of environmental stochasticity in relation to an organism's behaviour. However, the basic point applies to all phenotypic traits, not just behaviours. In this section we expound on the point and consider its implications for agential thinking of type 2.

Consider a simple example (inspired by Lewontin and Cohen 1969). A population is evolving in a temporally varying environment. There are two possible environmental states, s_1 and s_2, which occur with equal probability in each time-period, and independence across periods.[13] Reproduction occurs once per time-period, followed by death; so generations do not overlap. There are two types in the population, *A* and *B*, whose per-capita reproductive outputs, or fitnesses, are shown in Table 8.2. Thus in state s_2, for example, each *B* type leaves an average of 1.6 surviving offspring. We assume that type is perfectly inherited, and that population size can increase without

[12] Frank and Slatkin (1990) recommend the term 'risk-aversion' for the former and 'bet-hedging' for the latter; however, the latter is often used to refer to one particular way of reducing variability in fitness, namely using randomized strategies. See Starrfelt and Kokko (2012) for discussion of the meaning of bet-hedging.

[13] This independence assumption makes things easier but is not essential to the logic of the argument.

Table 8.2. Per-capita reproductive output of two types

	s_1	s_2
A	1.1	1.1
B	1.6	0.6

bound. So the cohort of *A*s grows by a factor of 1.1 in each state, while the cohort of *B*s grows by a factor of 1.6 in s_1 but only 0.6 in s_2.

Though *A* and *B* represent abstract phenotypes, rather than actions or behaviours, it is natural to think of them as 'safe' and 'risky', respectively. The *A* type ensures a constant, modest payoff across the two environmental states, while the *B* type does well in one state but badly in the other, so it takes a gamble. We can conceptualize this as a decision-theoretic problem if we want. However, note that here the 'choice' is performed by natural selection acting over generations, not by an individual organism within its lifetime; thus this is agential thinking (type 2). We can then ask whether natural selection will choose *A* or *B*.

Since states s_1 and s_2 are equi-probable, the expected reproductive output of both types is the same, namely 1.1 offspring per-capita—since $\frac{1}{2}(1.6) + \frac{1}{2}(0.6) = 1.1$. Thus it is tempting to conclude that the two types are selectively equivalent. For suppose we start with an equal number N_0 of *A*s and *B*s in generation zero. Then, after t generations the number of *A*s will be exactly $N_0(1.1)^t$, while the number of *B*s is a random variable with expected value $N_0(1.1)^t$. So surely neither type has an evolutionary advantage?

However, this does not follow, as has long been known. When the environmental state fluctuates over time, it is the geometric mean of reproductive output across states, not the arithmetic mean (or expectation), that is relevant to determining long-run success.[14] The arithmetic mean would only be relevant if the individuals added to the population in each generation did not go on to reproduce themselves, that is, if the expected increase in numbers between generations t and $t + 1$ were independent of population size N_t. But this is not how biological reproduction works. Rather, it is a multiplicative process in which the number of individuals added between t and $t + 1$ is a multiple of N_t. Simple limit calculations show that in the long-run, the type with the highest geometric mean will dominate the population with probability 1.[15]

In our example in Table 8.2, the geometric mean of type *A* is 1.1, while that of type *B* is $\sqrt{1.6} \cdot \sqrt{0.6}$, which is approximately equal to 0.97. Thus the *A* type has the advantage,

[14] The geometric mean of a random variable X, which takes values $X_1, \ldots, X_n$ with probabilities $p_1, \ldots, p_n$, is $\prod_{i=1}^{n} X_i^{p_i}$, while the arithmetic mean is $\sum_{i=1}^{n} X_i \cdot p_i$.

[15] For good expositions of the mathematics here, see Lewontin and Cohen (1969), Robson and Samuelson (2010), or Proulx and Day (2001), Appendix A.

and in the limit will grow without bound while the B type goes extinct. (After t generations, the number of Bs will be approximately $(0.97)^t$, which tends towards 0 as $t \to \infty$.) Thus natural selection acts in a risk-averse way, preferring the safer type A over the riskier type B, despite their having the same expected reproductive output. In essence, the A type is at an advantage because of its lower variance in output. Recall that if two random variables have the same expectation (or arithmetic mean), the one with the higher variance will have a lower geometric mean.

One easy way to see why the geometric not the arithmetic mean is the relevant quantity here is to modify the example slightly. Suppose that the B type earns payoffs of 2.2 and 0 in states s_1 and s_2, respectively, rather than 1.6 and 0.6. This leaves the expected payoff the same, at 1.1 offspring per-capita, but reduces the geometric mean payoff to zero. So it is still true that if we start with equal numbers of As and Bs in generation zero, then at any time t the expected number of As and Bs will be the same. But now it is obvious that the B type is doomed, since the whole cohort of Bs will go extinct as soon as the unfavourable state s_2 occurs for the first time; after ten generations, the probability that the B type is extinct exceeds 99.9%. This is a dramatic way of making a general point: in a fluctuating environment, a type that does very badly in some states pays a selective penalty that is missed if we focus on the expectation across states.[16]

The moral here is that natural selection 'cares' about the variance in reproductive output, not just the expectation, so displays a kind of intrinsic aversion to risk. As stressed in section 8.3.2, this refers to variability in the reproductive output of a *type*, not an individual. If there is variability in the output of individuals of the same type, this does not constitute a selective penalty, as in a large population it averages out. To illustrate, consider again the A type in Table 8.2. Its per-capita output of 1.1 in state s_1 could be realized in many ways. For example, perhaps $\frac{11}{20}$ of the As in state s_1 always leave 2 offspring while $\frac{9}{20}$ are killed by predators and leave none; or perhaps $\frac{11}{50}$ of the As leave 5 offspring while $\frac{39}{50}$ leave none. The particular realization makes no difference, as both lead to the same per-capita output of 1.1. It is the between-generation variability in the success of the type, not the within-generation variation among individuals of the type, that natural selection penalizes.

In biology, the former sort of variability is known as 'environmental' and the latter as 'demographic'. This corresponds closely to the economists' distinction between aggregate and idiosyncratic risk discussed previously. In essence, where the environment varies over time this gives rise to aggregate risk, as all individuals in a generation find themselves in the same environmental state. Where there is individual variability at a time, for example, because of predation which strikes at random, this gives rise to idiosyncratic risk. In practice, both sorts of risk may be found at once; see section 8.5.

[16] See Robson and Samuleson (2010) for a very clear explanation of this point.

8.4.1 Mother nature as a rational agent?

Recall agential thinking (type 2), the metaphorical assimilation of the process of natural selection to a rational agent's choice between alternatives. In chapter 1 we argued that agential thinking of this 'mother nature' sort works in simple cases but can potentially mislead. What about when the environment varies stochastically over time?

At first blush this seems to be a case where agential thinking (type 2) fails. For the paradigm of rational choice under risk is EU theory, which as we have seen involves taking expectations across states of the world, that is, evaluating by arithmetic mean utility. But natural selection 'prefers' the type with the highest geometric mean fitness across environmental states, as we have seen. Thus natural selection appears to use a quite different evaluation criterion from the one that a rational agent would use.

I think this is partially correct. Certainly it illustrates one way in which agential thinking (type 2) can lead astray if applied naively. In the foregoing example, if one assumes that natural selection's 'preferences' must accord with those of a rational agent aiming to maximize their reproductive success, who evaluates uncertain prospects in accordance with EU theory, one could easily conclude that selection will not discriminate between types *A* and *B*, given that their expected reproductive output is the same; but as we know, this is not true. This appears to reinforce our previous lesson, that agential thinking (type 2) is a heuristic of dubious value.

However, two qualifications are necessary here. The first is a simple mathematical point. If we order a set of random variables by geometric mean, this is equivalent to ordering them by the expected value of their logarithm. Therefore in a variable environment, in which a type's geometric mean fitness determines its evolutionary fate, the metaphor of natural selection as rational agent can in fact be salvaged. 'Mother nature' does evaluate alternatives in accordance with expected utility, just like a rational agent; it is just that her utility function is not reproductive output itself but rather its logarithm. Thus in our example, nature prefers the safe *A* type, which yields 1.1 offspring for sure, over the risky *B* type which yields expected number of offspring of 1.1, just as an EU-maximizer with a logarithmic utility function would prefer \$1.1 for sure over a risky prospect with an expected monetary value of \$1.1 (given that $\log(x)$ is concave). Indeed, if number of offspring is replaced by wealth, mother nature behaves just like the rational investor described by Bernoulli (1738), whose utility function is logarithmic in their wealth.

The correct moral, then, is that agential thinking (type 2) can in principle be used to understand how natural selection works in a stochastic environment, as long as we take care to specify mother nature's 'goal', that is, utility function, suitably. Defining utility as log(reproductive output) resolves the apparent mismatch between the evaluation criterion used by natural selection and by the rational agent. Whether agential thinking is *useful* in this case is a different matter; arguably it offers little heuristic guidance, since we need to do the evolutionary analysis in order to determine nature's utility function. But it is at least formally correct, in the sense that evolution can be represented as 'choosing' between types in accordance with the expected value of some quantity. Our second qualification, in the next section, reinforces this moral but for a different reason.

8.4.2 Long-run versus short-run measures of success

The geometric mean criterion is about the long-run evolutionary outcome; it derives from considering what happens in the limit. For this reason, some evolutionists have argued that short-term fitness measures are also needed to achieve a full understanding of how evolution works in variable environments.[17] One such measure is the expected reproductive output of a type relative to the population average, or expected relative fitness for short.[18] The sign of this quantity determines whether the type's frequency is expected to increase or decrease in the next generation, which is a natural thing to be interested in when the change in frequency is a random variable.

To illustrate, consider again the *A* and *B* types in the previous section's example, with reproductive fitnesses given in Table 8.2. Let p denote the frequency of the *A* type in generation t (censused at the time of conception, before the environmental state that generation is determined). The change in frequency over a single generation, Δp, is thus a random variable whose value depends on which environmental state occurs. It is easy to show that:

$$\mathbf{E}[\Delta p] > 0 \quad \text{if and only if} \quad \mathbf{E}\left[\frac{w_A}{\overline{w}}\right] > 1 \tag{8.1}$$

where $\mathbf{E}$ denotes expectation, w_A is the per-capita reproductive output of the *A*-type, and $\overline{w}$ is the population average reproductive output. Note that w_A and $\overline{w}$ are both random variables. We can see from Table 8.2 that w_A is always equal to 1.1, while $\overline{w}$ takes value $(1.6 - 0.5p)$ if state s_1 occurs and $(0.6 + 0.5p)$ if state s_2 occurs.

As before, we assume that states s_1 and s_2 occur with equal probability, and that the *A* and *B* types begin with equal frequency, that is, $p = \frac{1}{2}$. This gives $\mathbf{E}[\frac{w_A}{\overline{w}}] \approx 1.05$ and $\mathbf{E}[\frac{w_B}{\overline{w}}] \approx 0.95$. Therefore by equation (8.1) we see that $\mathbf{E}[\Delta p] > 0$, that is, the expected frequency of *A* will increase over a single generation while that of *B* will decrease. By this short-term criterion, therefore, the safer *A* type has an advantage over the riskier *B* type over a single generation. Note the contrast here with the geometric mean criterion, which is essentially about an advantage that accrues over multiple generations.

The intuitive explanation of this is as follows. The *A* type does equally well in both states, but the *B* type does much better in state s_1 than s_2. In virtue of this, the average population fitness $\overline{w}$ is higher in s_2 than state s_1. Thus although the expected *absolute* fitness of the two types, across environmental states, is the same, the expected fitness

[17] See for example Lande (2007) and Proulx and Day (2001).

[18] Importantly, 'relative' here means relative to the population average, not relative to the other type. This ambiguity in the meaning of 'relative fitness' has been noted by Rice (2008).

of the A type *relative to the population average* is greater. In effect, the relative fitness of the B type is diluted, in expectation, by the fact that in the environmental state where it does well the population as a whole does well too. The A type thus has the advantage due to its more consistent performance, that is, its lower variance across states.

This highlights another respect in which evolution behaves in a risk-averse way. In general terms, the point is that there is a concave relationship between the absolute and relative fitness of a type (Frank and Slatkin 1990). To see this, consider a single environmental state. The absolute fitness of type A is w_A while its relative fitness is $w_A/(pw_A + (1-p)w_B)$, which is a concave function of w_A. That is, if the absolute fitness of type A is increased by a single unit, its relative fitness increases by less than a unit; similarly for the B type. Therefore, when we consider multiple environmental states, types with a high variance in absolute fitness, such as the B type, will suffer a reduction in their expected relative fitness, and thus in their expected frequency in the next generation.

In our example the A type, thanks its lower variance in absolute fitness, does better by the criterion of geometric mean fitness, and also by the criterion of expected relative fitness (starting from $p = \frac{1}{2}$). However, in general, these two criteria are not ordinally equivalent, as Autzen (2017) has stressed. One way to see this is to note that expected relative fitness is a frequency-dependent quantity while geometric mean fitness is not. By equation (8.1), this means that $\mathbf{E}[\Delta p]$ may be positive for some values of p and negative for others. Thus the ordering of the types by their expected relative fitness may be inverted as the population evolves. In general, if $\mathbf{E}[\Delta p] > 0$ for all values of p, then the A type must have the higher geometric mean and so will evolve in the long-run, but not vice-versa.[19]

What does all this show? I want to stress two points. First, if one uses expected relative fitness as the criterion of evolutionary success, in a variable environment scenario, this points to a further respect in which natural selection acts in a risk-averse way, distinct from considerations of geometric mean. Second, this again allows agential thinking (type 2) to be salvaged. Mother nature does choose between the competing types according to expected utility, but in this case her utility function is not reproductive output itself, nor its logarithm, but rather reproductive output divided by the population average. By defining utility this way, we can again ensure that nature's short-term preference coincides with that of a rational agent who conforms to EU theory.

The issue of whether $\mathbf{E}[\Delta p]$ over a single generation is a quantity that evolutionary biologists should care about for its own sake is of course open to debate. Surely what matters is evolutionary success in the long-run? This is a complex issue to which I cannot do justice; the answer partly depends on what biological question one wants the fitness concept to answer.[20] The point here is simply that *if* one is interested

[19] See Okasha (forthcoming) for a proof of this point.
[20] See Lande (2007) and Starrfelt and Kokko (2012) for discussion of this issue.

in $\mathbf{E}[\Delta p]$, then expected relative fitness is the 'natural' criterion of evolutionary advantage; and this allows the evolution/rational choice parallel to be restored.

8.4.3 A parallel with finance

A parallel with financial economics deserves brief mention. In portfolio choice, an investor must decide how to allocate their wealth among different financial assets repeatedly over time, with the profits from one period being re-invested in the next. The return on each asset is a random variable, subject to a known probability distribution; in the simplest case, this distribution is assumed equal in all periods with independence across periods. This is closely analogous to population growth in a stochastic environment, as it also concerns a quantity that grows multiplicatively, with a random growth rate. Unsurprisingly, the geometric mean criterion arises in this context too, and has been given a similar justification (Kelley 1956). An investor who, in each period, chooses a portfolio that maximizes geometric mean payoff will end up with more wealth in the long-run than any other investment strategy, with probability one.

Samuelson (1971) famously criticized the idea that the geometric mean criterion is rationally compelling in portfolio choice. He observed that it conflicts with expected utility maximization unless the investor's utility function happens to be logarithmic in their wealth, which it may well not be. He also observed that the criterion's justification is essentially long-run, but in the real world all sequences of investments are finite; and in a finite sequence of any length, choosing so as to maximize the geometric mean need not bring higher expected wealth, or utility. Both of these points have corollaries in the evolutionary case. In our example of evolutionary competition between the A and B types in a variable environment, in which A has the advantage by the geometric mean criterion, natural selection can be validly analogized to rational choice, but *only* if utility is defined as log(reproductive output). And it is also true that, although type A has a long-run advantage, after finitely many generations t, the expected numbers of As and Bs are identical, if they are initially equally abundant.

Despite this, the justification for focusing on the geometric mean is arguably stronger in evolutionary biology than in finance (Brennan and Lo 2011). For interesting biological adaptations typically evolve over many thousands of generations; so even if some environmental states occur only rarely, say with probability $\frac{1}{20}$ per generation, the geometric mean fitness will be a good predictor of the likely fate of a genotype.

8.5 Evolution of Irrationality?

In section 8.3, we discussed organisms' choices between options with uncertain rewards, while in section 8.4 we discussed evolution's 'choices' between traits in uncertain environments. In both cases, we asked whether the choices in question conform to the rationality principles of EU theory, and thus whether agential thinking of types 1 and 2, respectively, can be applied. In this section we link the two issues by returning to the questions: what sort of risk preferences should we expect organisms to evolve, given how natural selection works in a risky environment? Will organisms' choices appear rational or not? Thus we return to agential thinking (type 1), that is, the organism-as-rational-agent idea, but in the light of our analysis of how natural selection 'chooses' in the face of risk.

Consider again an organism facing a choice between two actions X and Y. As before, action X yields lifetime reproductive output of 5 offspring for sure, while Y yields 1 or 9 offspring with equal probability. We saw that if the risk is purely idiosyncratic, the organism should evolve to be indifferent between X and Y: both are equally good. Similarly, if a rational agent obeys EU maximization, and their utility function is equal to (or linear in) their reproductive output, they will be indifferent between X and Y, since $\text{EU}(X) = \text{EU}(Y)$. So with idiosyncratic risk, evolution can induce an organism to make the biologically optimal choice by equipping them with this utility function and having them choose in accordance with EU theory.

What if the risk is purely aggregate? As we know, X is then evolutionarily superior to Y by the geometric mean principle. Then, if we set utility = log(reproductive output), we can again make the EU-maximizing action coincide with the biologically optimal action. With this utility function, a rational agent's evaluation of the options will be $\text{EU}(A) = \log 5$, and $\text{EU}(B) = \frac{1}{2}(\log 1) + \frac{1}{2}(\log 9)$ which is less than log 5; so the agent will prefer A. This follows from the fact that ordering random variables by geometric mean is equivalent to ordering them by the expected value of their logarithm. Thus by equating utility with the logarithm of reproductive output, rationality and biological optimality will again coincide.

Taken together, these points suggest that in both the idiosyncratic and aggregate risk cases, evolution will lead organisms to have preferences (between lotteries over offspring) that satisfy EU maximization. When the risk is idiosyncratic, evolution will lead organisms to value additional offspring linearly; when it is aggregate, it will lead them to value additional offspring in inverse proportion to how many they already have. In both cases, if an organism's choices accord with EU maximization, they will be biologically optimal, so agential thinking (type 1) is applicable: the organism can be treated as if trying to achieve a goal. Thus where risk is concerned, rationality and adaptiveness can seemingly be reconciled, as long as the utility function is appropriately defined.

The situation does not change fundamentally if we consider an organism choosing between actions that yield uncertain amounts of food or energy, rather than offspring directly. Suppose that $w(x)$ is the function mapping food intake (x) per period onto additional number of offspring; this function may have any shape. Consider the organism's preferences between lotteries over food rewards, such as 'nine or one units of food with equal probability'; we may assume that each available action induces such a lottery. If the risk is purely idiosyncratic, then to be biologically optimal the organism should choose like an EU maximizer whose utility function for x is linear in $w(x)$, while if the risk is purely aggregate, their utility function should be linear in $\log w(x)$. In both cases, therefore, rational risk preference seems to evolve, so the organism-as-rational-agent heuristic works fine.

This conclusion is correct in so far as it goes; however, it has two important limitations. First, it only deals with the limiting cases of purely idiosyncratic and purely aggregate risk. When both sorts of risk are present simultaneously, matters are different. Second, we have not considered mixed strategies, which involve randomizing over actions, but these are crucial in considering evolution in variable environments. These points are addressed in turn in the next sections.

8.5.1 A biological Allais paradox

Suppose organisms in an evolving population encounter two different choice situations in their lifetime. In situation 1, they must choose between action a, which yields 1 offspring for sure, and action b, which yields either 0, 1, or 5 offspring with probabilities of $\frac{2}{10}$, $\frac{7}{10}$, and $\frac{1}{10}$, respectively. In situation 2, they must choose between action c, which yields 0 or 1 offspring with probabilities $\frac{7}{10}$ and $\frac{3}{10}$, and action d, which yields 0 or 5 offspring with probabilities $\frac{9}{10}$ and $\frac{1}{10}$. Note that these payoffs correspond exactly to the version of the Allais paradox described in section 8.2, with 1 offspring = \$1 million, as is clear from Table 8.3.

Suppose that each organism encounters situations 1 and 2 exactly once in its lifetime, in succession; its total reproductive output is the sum of its payoff in each situation. Further, let us stipulate that in situation 1 the risk is purely aggregate, but in situation 2 it is purely idiosyncratic. That is, in situation 1 the probabilities are perfectly correlated across organisms choosing the same action (with independence across generations), while in situation 2 an organism's choice of action, in any generation, induces a lottery over the payoffs that is independent of the lottery faced by other organisms making the same choice.

An organism's type specifies which action it performs in each situation. Thus there are four types in the population: (ac), (ad), (bc), (bd). We assume that the population is infinite, that type breeds true, and that no non-selective forces are present. Our question is then: which type will win out in evolutionary competition?

To address this, we need to consider the per-capita reproductive output of each type in a generation, which is a random variable, and then compute the long-run

Table 8.3. A biological Allais paradox

		$\frac{2}{10}$	$\frac{1}{10}$	$\frac{7}{10}$
Situation 1	a	1 offspring	1 offspring	1 offspring
(aggregate)	b	0	5 offspring	1 offspring
Situation 2	c	1 offspring	1 offspring	0
(idiosyncratic)	d	0	5 offspring	0

Table 8.4. Reproductive output of each type

Type	(Rep. output, probability)	Geometric mean
(ac)	(1.3, 1)	1.3
(ad)	(1.5, 1)	1.5
(bc)	$(0.3, \frac{2}{10}; 5.3, \frac{1}{10}; 1.3, \frac{7}{10})$	1.12
(bd)	$(0.5, \frac{2}{10}; 5.5, \frac{1}{10}; 1.5, \frac{7}{10})$	1.37

growth rates by the geometric mean criterion. These are given in Table 8.4. To illustrate the calculations, consider first the (ac) type. In situation 1, each (ac) organism gets 1 offspring. In situation 2, each (ac) organism faces an independent lottery with expected payoff of 0.3 offspring. Since the population is infinite and the risk idiosyncratic, the average payoff to the (ac) type in situation 2 will be 0.3, by the law of large numbers. Adding over both situations, the per-capita reproductive output of the (ac) type will be 1.3 offspring with probability 1; thus the geometric mean is 1.3.

By contrast, consider the (bd) type. In situation 1, a single lottery determines whether each (bd) organism gets 0, 5, or 1 offspring, with probabilities of $\frac{2}{10}$, $\frac{1}{10}$, and $\frac{7}{10}$. In situation 2, each (bd) organism faces an independent lottery with expected payoff of 0.5 offspring; thus by the law of large numbers, the average payoff to the (bd) type in situation 2 will be 0.5. Adding over situations, the per-capita reproductive output of the (bd) type, in any generation, will be 0.5 with probability $\frac{2}{10}$, 5.5 with probability $\frac{1}{10}$, and 1.5 with probability $\frac{7}{10}$; the geometric mean works out as approximately 1.37.

As Table 8.4 shows, the (ad) type has the highest geometric mean fitness out of our four types, so in the long-run it will dominate the population. This result is striking, since an organism of the (ad) type exhibits the Allais preferences, which as we know are incompatible with maximization of expected utility however, the utility function is chosen. (More precisely, there exists no utility function, whose sole argument is reproductive output, such that the expected utility of (ad) is highest.) So when the (ad) type takes over the population, as it must do, an observer studying the risk-preferences of evolved organisms will find that they violate EU maximization. That is, the organisms exhibit irrational choice behaviour, so they cannot be regarded as akin to rational agents trying to achieve a goal.

What drives this unusual result? Essentially, it arises because of the particular combination of aggregate and idiosyncratic risk. In situation 1, where the risk is aggregate, action b is selectively disadvantageous since it yields a zero payoff in one state of the world; this is in spite of its having a higher expected payoff than a. In situation 2, where the risk is idiosyncratic, action d is superior by virtue of its higher expected payoff. Given that each organism faces both situations within its lifetime, it is thus intuitive that the (ad) combination should do best overall; this is borne out in the geometric mean calculations.

It follows that our two modes of agential thinking diverge, in this example. Nature 'chooses' between the types rationally, according to the geometric mean criterion, that is, expected log fitness; but this results in organisms who themselves make irrational choices, that is, ones which cannot be represented as maximizing expected utility however, their utility function is defined. Thus agential thinking (type 2)—the metaphorical assimilation of the process of natural selection to a rational agent's choice—can in principle apply, but agential thinking (type 1) cannot. In this case, therefore, adaptiveness and rationality part ways, in the sense that biological optimality leads to economic irrationality.

The fact that the Allais preferences evolve in our example is striking, but we should not read too much into it. For the particular combination of aggregate and idiosyncratic risk was of course jury-rigged to yield the desired result. We could just have easily posited that the risk in situation 1 was idiosyncratic and in situation 2 aggregate, and a different result would have obtained. Thus our example is not intended as a serious candidate for explaining how Allais preferences might have evolved in humans or animals (if they did), but rather serves an illustrative purpose. It shows that when arbitrary combinations of aggregate and idiosyncratic risk are considered, irrational risk preferences may evolve.

An example of Robson (1996) makes a related point, though without reference to the Allais paradox. Robson considers a choice between lottery *A*, which yields either 1 or 9 offspring with equal probability, and *B*, which yields either 1 or 8.5 offspring with equal probability. In *A* the risk is aggregate, that is, a single coin is flipped for all population members, where in *B* it is idiosyncratic, that is, each individual flips a separate coin. From the perspective of any single individual, an action that leads to lottery *A* is preferable to one that leads to *B*—since *A* stochastically dominates *B*.[21] However, evolution will favour organisms that choose action *B*, since they do better by the geometric mean criterion.[22] Since an EU-maximizing agent will never choose a stochastically dominated lottery, again it seems that evolution produces irrational risk preference.

It is an open question whether the empirically documented violations of EU-maximization have a biological basis, and if so, how they evolved. That is an empirical issue, but here I want to stress a philosophical moral. In thinking about adaptive behaviour in the face of risk, it is tempting to assume that the organism-as-rational-agent heuristic can apply, as long as organisms' preferences have been optimized by evolution. However, appearances can deceive. We have seen that if the risk in question is purely aggregate or purely idiosyncratic the heuristic will indeed work, given a suitable choice of utility function. But where both sorts of risk are present together,

[21] This means that for any payoff level, the probability of getting that level or more is never lower under *A* than under *B*.

[22] The per-capita reproductive output of the *A* type is 4.5, while of the *B* type is 1 or 8.5 with equal probability; and $4.5 > \sqrt{8.5}$.

this is no longer the case. This points to a logical limit on the use of agential thinking (type 1) to understand evolved behaviour.

8.6 Bet-hedging and Mixed Strategies

A different challenge to agential thinking (type 1), in situations involving risk, arises from mixed strategies. This relates to the well-known concept of bet-hedging in biology.[23] Consider the following example (inspired by Cooper and Kaplan 1982). Each year a species' environment is either cold or warm with probabilities q and $(1 - q)$, respectively, and independence across years. Organisms live for a year then die, and make a one-off choice in their lifetime between two actions: X and Y. Action X leads to three offspring in a cold year but zero in a warm year; action Y leads to zero offspring in a cold year but four in a warm year (Table 8.5).

Clearly, a genotype coding for action X will go extinct in the first warm year, as will one for Y in the first cold year; both have geometric mean fitness of zero. However, a genotype that randomizes between X and Y, with independence across individuals, can do better than this—for in any year some individuals with the genotype will survive, thus allowing the lineage to avoid extinction. When a genotype codes for a randomized strategy, we can imagine each individual with the genotype flipping a (possibly biased) coin to decide whether to do X or Y. The use of randomized (mixed) strategies in response to environmental risk is one meaning of the term 'bet-hedging' in biology, and is the focus here.[24] Intuitively it is advantageous because it 'spreads the risk', or avoids putting all the eggs in one basket, and is akin to the investment strategy of holding a diversified portfolio.

The basic logic behind bet-hedging derives from the geometric mean principle, and the consequent selection against types with a high variance in reproductive success across environmental states (and thus generations). Essentially, a genotype that randomizes between X and Y reduces the correlation in fitness among individuals of that genotype in any generation; and this has the effect of levelling out the genotype's

Table 8.5. Cold and warm environments

	Cold	Warm
Action X	3	0
Action Y	0	4

[23] See Seger and Brockmann (1987) or Starrfelt and Kokko (2012) for good introductions to bet-hedging.

[24] Some authors distinguish between diversified bet-hedging, which involves randomization, and conservative bet-hedging, which simply involves using a strategy with low variance in fitness. See Starrfelt and Kokko (2012) for discussion.

average reproductive success between generations. Note that a bet-hedging genotype only has an advantage when there is environmental variability (hence aggregate risk). In a fixed environment, a mixed strategy has no evolutionary advantage.

In the cold/warm example, any mixed-strategy genotype does better than the pure strategy of always doing X or always doing Y. But the *optimal* genotype, denoted G^*, randomizes between X and Y in a particular ratio, namely $q:1-q$, as this leads to maximal geometric mean fitness.[25] Randomizing in this way is sometimes called 'probability matching', since the fraction of individuals using X, in the optimal mixture, equals the probability of a cold year. Thus if cold and warm years are equally probable, that is, $q=\frac{1}{2}$, then the optimal genotype G^* has each individual flip a separate fair coin to decide between X and Y; the genotype then receives an average payoff of 1.5 offspring in a cold year and 2 in a warm year (assuming a large population). Note that the expected payoff to G^* is 1.75, less than that of a genotype coding for the pure action Y, but the former has a higher geometric mean thanks to its lower variance.

Bet-hedging is an interesting topic in its own right, but here my concern is with its implications for the organism-as-rational-agent idea. One natural thought is that it conflicts with this idea, since it leads to a pattern of preference that no rational agent should have. To see why, note that an EU-maximizing agent will never exhibit a strict preference for randomization in a decision-theoretic context. Given two options X and Y, the lottery 'X with probability p, Y with $(1-p)$' can never have expected utility strictly greater than that of X, and than that of Y. But in the foregoing example, a genotype that randomizes over X and Y is strictly superior to one that always produces X or always Y, as we have seen. On the face of it, therefore, adaptiveness and rationality part ways here. If a mixed strategy evolves to fixation, in the foregoing example, the population will be composed of individuals who strictly prefer to mix between X and Y, and whose risk preferences thus cannot be represented as EU-maximizing.

It is important not to confuse having a strict preference for randomization with being risk-prone, as for example when an agent prefers a fair bet on \$0 or \$10 to the definite receipt of \$5. The latter is fully compatible with EU maximization; it simply means that the agent's utility function for money is convex in the relevant region, as we know. By contrast, a strict preference for randomization would mean strictly preferring a bet on \$0 or \$10 to the certain receipt of \$0, and also preferring the bet to the certain receipt of \$10. Such a preference is both intuitively irrational and in direct contravention of EU theory.

[25] To see this, consider an arbitrary mixed strategy of the form 'X with probability p, Y with probability $(1-p)$'. This has geometric mean fitness $G = 3p^q \cdot 4(1-p)^{(1-q)}$. Since maximizing G is equivalent to maximizing its logarithm, we may simply maximize $\ln G = q \ln 3p + (1-q) \ln 4(1-p)$ with respect to p and solve, which yields $p = q$.

There is some evidence that human decision-makers do actually exhibit a strict preference for randomization in certain contexts.[26] (This is closely related to the phenomenon of 'ambiguity aversion' discovered by Ellsberg (1961), in which people prefer to bet on events with known rather than unknown probabilities, a well-known violation of EU theory.) It is possible that the evolutionary origin of such preferences lies in the adaptive advantage of mixed strategies in uncertain environments, an argument made by Brennan and Lo (2011). The suggestion might be thought implausible, given that clearcut empirical examples of randomizing behaviour in biology are sparse.[27] But rather than speculating about this, here I want to further probe how mixed strategies relate to agential thinking (type 1).

We argued previously that where organisms face purely aggregate risk, their evolved risk preferences should satisfy EU theory with utility equal to log(reproductive success). However, this argument was made in the context of pure strategies (where it holds good). Where mixed strategies are considered matters are different, and indeed here it seems that evolution again leads to irrationality. Nature 'chooses' a strategy that leads individual organisms who bear the strategy to exhibit anomalous choice behaviour, such as preferring to randomize between actions.

One possible response to this argument is that we are interpreting randomization too literally. After all, what really matters to the success of the optimal genotype G^*, in our bet-hedging example, is just that in any year on average half the individuals with the genotype perform X and half perform Y (in the $q = \frac{1}{2}$ case.) This might be achieved by each individual actually flipping a coin, or doing something equivalent such as conditioning their choice on a suitable random event, but not necessarily. Alternatively, the mixing could be implemented at the level of the genotype. That is, the genotype could simply produce two types of individual in equal proportion, one with a fixed preference for action X and the other for Y, that is, a polymorphism. So no individual with the genotype need actually prefer to randomize.

This is certainly true, but does it really help? After all, individuals who choose action X receive 0 or 3 offspring with equal probability, so have expected fitness of 1.5; those who choose Y receive 0 or 4 with equal probability, so have expected fitness of 2. So if the optimal genotype G^* is implemented via polymorphism rather than individual randomization, it still seems to lead to irrationality: half the individuals with the genotype choose an action that is (stochastically) dominated, and are thus violating EU maximization. If the genotype G^* fixes in the population, an observer will still see individuals who appear to make an irrational choice, and whose behaviour thus resists interpretation in agential terms. So we have simply traded one sort of irrationality—preferring to randomize—for another—preferring a stochastically dominated option.

[26] For discussion of the empirical data, see Agranov and Ortoleva (2017).

[27] Indeed empirical examples of bet-hedging in general are relatively sparse, and what examples there are do not appear to involve individuals actually engaging in randomization. See Childs *et al.* (2010) for a recent survey.

However, there is a possible comeback to this argument. It derives from a point made by Grafen (1999) in response to Cooper and Kaplan (1982), who argued that when a bet-hedging genotype evolves, some individuals with the genotype are 'altruistic'. (Their argument does not depend on whether the bet-hedging involves individual randomization or polymorphism.) In terms of our example, the Cooper and Kaplan idea is that if the optimal genotype G^* evolves, then an individual who performs action X under G^* is sacrificing their expected fitness for the long-term good of their genotype, or lineage. Switching to Y would lead to a gain in expected fitness for any such individual, but at a cost to the genotype as a whole.

In response, Grafen argues that this uses the wrong criterion of individual advantage. Instead, we should consider the expectation of an individual's fitness *relative to the population average*, he argues. Suppose the optimal genotype G^* is fixed in the population (again, in the $q = \frac{1}{2}$ case). Then, the average population fitness $\overline{w}$ is 1.5 in the cold state, and 2 in the warm state. Thus the expected relative fitness of an individual who chooses action X is $\frac{1}{2} \cdot \frac{3}{1.5} = 1$, while that of one who chooses Y is $\frac{1}{2} \cdot \frac{4}{2} = 1$. Both individuals do equally well by this criterion, as does one who randomizes between the two actions. So at evolutionary equilibrium no individual is behaving altruistically, nor choosing a stochastically dominated action.

The debate between Grafen and Cooper and Kaplan is about altruism; however, it also has implications for rationality (as Grafen notes). For Grafen's point suggests that, by suitably re-interpreting what we take an individual's goal or utility function to be, the behaviours that result from a bet-hedging strategy can in fact be reconciled with individual rationality. For although it is clear that when bet-hedging evolves, individuals do not maximize the expectation of a utility function whose sole argument is their own number of offspring, Grafen's argument suggests that this no longer holds true when more complicated utility functions are permitted. I explore this next.

8.6.1 The McNamara/Robson model

To focus the issue, we consider a simplified version of an evolutionary model, due independently to McNamara (1995) and Robson (1996), which is in the background of Grafen's argument. The model generalizes the cold/warm environment example of Table 8.5, while also incorporating idiosyncratic risk. There is a finite set of environmental states S, one of which occurs each year with independence across years; $q(s)$ is the probability distribution on S. There is a finite set A of pure actions available to an individual. In state $s \in S$, an individual who performs action $a \in A$ receives a lottery over the possible number of offspring with expected value $w_a(s)$; this is an expectation over idiosyncratic risk. The limiting case of purely idiosyncratic risk arises if S is a singleton, and of purely aggregate risk if each action in each state yields a certain number of offspring rather than a lottery, that is, $w_a(s)$ denotes an actual rather than an expected number of offspring.

Each individual has a strategy, which is a probability distribution over actions. The set of possible strategies, denoted T, is the set of all such distributions. Any two individuals with the same strategy randomize identically over the actions, but independently of each other. A pure strategy is one in which all the probability is placed on one action. The question is then: which strategy will evolve?

In a given state s, the payoff to a strategy R in which each action $a \in A$ is played with probability $p(a)$ is:

$$w_R(s) = \sum_A p(a) \cdot w_a(s)$$

Note that this is an expectation across actions, according to their probability of being played under strategy R. Since individuals using the same strategy randomize independently, in a large enough population $w_R(s)$ equals the average payoff of the cohort of individuals using strategy R in a given year.

Since the environment varies over time, a strategy's long-run rate of increase depends on its geometric mean payoff. An optimal strategy R^* is one that maximizes the geometric mean, hence maximizes:

$$\sum_S q(s) \cdot \log(w_R(s)) \tag{8.2}$$

This is the expectation across environmental states of the logarithm of a strategy's payoff. Therefore, R^* is optimal if and only if $\sum_S q(s) \cdot \log(w_{R^*}(s)) \geq \sum_S q(s) \cdot \log(w_R(s))$ for all strategies $R \in T$. Note that there may be more than one optimal strategy.

This analysis confirms two of our previous claims. First, when the risk is purely idiosyncratic, evolution will lead organisms to behave like EU maximizers with utility = reproductive output. In this case S is a singleton, so an optimal strategy simply maximizes $\log(w_R)$, which is equivalent to maximizing w_R itself, that is, the strategy's payoff. Such a strategy will place all the probability on actions $a \in A$ that maximize expected number of offspring $w_a(s)$; so an organism with the strategy will always choose actions that maximize $w_a(s)$. Second, when the risk is purely aggregate, if only pure strategies are considered, then evolution will lead organisms to behave like EU maximizers with utility = log(reproductive output). In this case, each $w_a(s)$ is an actual rather than an expected number, as is the payoff to a pure strategy $w_R(s)$. Thus by equation (8.2), an organism using an optimal pure strategy will always choose an action that maximizes the expectation, across states, of $\log w_a(s)$. Hence organisms again exhibit rational choices.

However, the interesting case is when mixed strategies are allowed, where rationality is more elusive. In this case, McNamara (1995) offers an alternative characterization of an optimal strategy, which is the basis of Grafen's point. He shows that an optimal strategy R^* must satisfy:

$$\sum_S q(s) \left(\frac{w_R(s)}{w_{R^*}(s)} \right) \leq 1 \quad \text{for all strategies } R \in T \tag{8.3}$$

Since equality obtains when $R = R^*$, it follows that an optimal strategy is one that maximizes the expectation, across environmental states, of a strategy's payoff relative to itself.

To understand this, consider again the cold/warm environment example of section 8.6, with $q(\text{cold}) = q(\text{warm}) = \frac{1}{2}$. We saw that the optimal strategy G^* then mixes equally between actions X and Y; its payoff $w_{G^*}(s)$ is 1.5 when s is cold and 2 when s is warm. To confirm that G^* satisfies condition (8.3), consider an arbitrary strategy R_p which mixes between X and Y with probabilities of p and $1-p$, earning payoff of $w_{R_p}(s) = 3p$ when s is cold and $4(1-p)$ when s is warm. Then, the expected payoff of R_p relative to G^* is:

$$\sum_S q(s)\left(\frac{w_{R_p}(s)}{w_{G^*}(s)}\right) = \frac{1}{2}\cdot\frac{3p}{1.5} + \frac{1}{2}\cdot\frac{4(1-p)}{2} = 1$$

hence equation (8.3) is satisfied. By contrast, consider the sub-optimal mixed strategy $R_{\frac{1}{4}}$, that is, which chooses X with probability $\frac{1}{4}$. The expected payoff of R_p relative to $R_{\frac{1}{4}}$ turns out to be $(4p+2)/3$, which is less than 1 when $p < \frac{1}{4}$, hence $R_{\frac{1}{4}}$ does not satisfy the optimality condition (8.3). It is easily confirmed that G^* is the unique strategy satisfying equation (8.3), in this example.

8.6.2 *Rationality restored?*

Let us return to the issue of whether individual organisms behave rationally or not, in a bet-hedging scenario. There are two possible takes on this issue. The first stems from the following observation of McNamara (1995) in relation to the foregoing model. An optimal strategy R^* does not maximize expected payoff across environmental states, as we know; rather it maximizes the geometric mean. However, McNamara notes that we can define a biased probability distribution on the states $q^*(s)$, such that an optimal strategy does maximize 'expected' payoff relative to $q^*(s)$; moreover, every action that gets positive probability in the optimal mixture receives the same 'expected' payoff. This biased distribution $q^*(s)$ shifts probability mass away from states where the optimal strategy performs well, onto states where it performs badly; it is defined by the relation $q^*(s) \propto \frac{q(s)}{w_{R^*}(s)}$.[28] We can see from condition (8.3) that relative to this biased distribution, an optimal strategy R^* does indeed receive 'expected' payoff no less than any other strategy R. To see that all actions receiving positive probability under R^* do equally well, note that if this were not so, R^*'s 'expected' payoff could be increased by moving probability onto an action that does better, in which case R^* would not be optimal.

To illustrate this point, consider again the cold/warm environment example with $q(\text{warm}) = q(\text{cold}) = \frac{1}{2}$. The unique optimal strategy G^* achieves a payoff of

[28] The constant of proportionality is $\frac{1}{K}$, where $K = \sum_S \frac{1}{w_{R^*}(s)}$; hence $q^*(s) = \frac{1}{K}\cdot\frac{q(s)}{w_{R^*}(s)}$. This ensures that $q^*(s)$ is a true probability distribution, that is, $\sum_S q^*(s) = 1$.

1.5 and 2 in the cold and warm states, respectively, as we know. Thus the biased probability distribution on the states is given by $q^*(\text{cold}) = \frac{4}{7}$ and $q^*(\text{warm}) = \frac{3}{7}$. It is easily verified that when we compute the expectation using q^*, strategy G^* maximizes 'expected' payoff, that is, no other strategy, pure or mixed, does better. Moreover, an individual who chooses action X under G^* receives 'expected' payoff of $\frac{4}{7} \cdot 3 = \frac{12}{7}$, while one who chooses Y receives $\frac{3}{7} \cdot 4 = \frac{12}{7}$. Therefore neither gains by switching.

What does this show? The natural interpretation, I suggest, is that when evolution fixes the optimal strategy in the population, in response to a fluctuating environment, this results in individual organisms choosing between actions in a non-rational fashion. It is as if individuals know the true probability distribution on the environmental states q, but deliberately use a different distribution to compute the 'expected' payoff of the possible actions. Indeed there is a clear similarity here with non-expected utility theory, the family of alternatives to standard EU theory mentioned in section 8.2, in which the decision weights diverge from the true probabilities. Non-EU theory enjoys some success as a description of humans' risk preferences but lacks a normative basis, that is, it models less than fully rational behaviour. It is striking that in this simple model, individuals with an optimal strategy are led to choose between actions in a manner similar to how non-EU maximizers choose. This too suggests a limit on agential thinking (type 1): here evolution leads organisms to behave not like rational agents, but like ones who deviate systematically from the rational ideal.

However, a different though compatible take comes from Grafen (1999) and Curry (2001). Grafen observes that at equilibrium in the foregoing model, every individual chooses an action that maximizes the expectation across states, relative to the true distribution q, of its payoff relative to the population average. To see this, note that when a strategy R is fixed in the population the average population fitness in state s is $w_R(s)$. From the characterization of optimality in equation (8.3), it follows that an optimal strategy R^*, when fixed, maximizes expected payoff relative to the population average. By reasoning identical to that used previously, every action that receives positive probability under R^* also maximizes expected relative fitness, when R^* is fixed.

It follows that individual rationality can be salvaged after all. Under the optimal strategy individuals do choose rationally; however, an individual's goal is not to maximize its number of offspring but rather to maximize its number relative to the population average, that is, to maximize its share of the total population's reproductive output, or fitness. More precisely, define the utility an individual gets from performing action a in state s as $w_a(s)/\overline{w}(s)$, where $\overline{w}(s)$ is average population fitness in state s. Then, when an optimal strategy is fixed in the population, each individual chooses between actions in accordance with EU maximization.

One consequence of this, stressed by Curry (2001), is that there is a hidden game-theoretic aspect to the evolution of bet-hedging strategies. An individual's relative fitness, in a given environmental state, depends in part on how well the population as a whole does in that state; and this depends on which actions other individuals in the population choose. Thus although an individual's absolute fitness $w_a(s)$ depends only on its own action a, its relative fitness depends on those of others. Thus although the problem was initially posed as a purely decision-theoretic one, it has an inherently strategic dimension.

Which of these two interpretations should we prefer? I see no need to choose between them. When the optimal strategy is fixed, we can either think of organisms as having the goal of maximizing their absolute number of offspring and using non-rational means to that end, that is, a non-EU evaluation criterion, or as having the goal of maximizing their fractional share of the population's reproductive output and using a rational means to that end, that is, evaluating by EU maximization. These are equivalent perspectives: there is no fact of the matter as to which is correct.

This duality of perspectives is not altogether surprising. It is a biological analogue of a moral familiar from rational choice theory, namely that apparent anomalies of choice can often be eliminated by changing the state-space. We saw in section 8.2 that violations of the independence axiom of EU theory can always be explained away by re-describing the outcomes of the actions. Something similar is true here. If the outcomes have the form 'leave x offspring', then an organism's choices over lotteries will violate EU maximization; however, if the outcomes are re-described as 'leave y% of the total population's reproductive output', then the organism's choices will appear rational. However, in this case, the re-description necessary to restore rationality is quite specific.

This is an instance of a general moral emphasized in previous chapters. Valid use of agential thinking (type 1) in biology, in which an evolved organism is treated as an agent with a goal towards which its phenotypic traits conduce, requires that we think carefully about what the 'goal' is supposed to be. Traits that are anomalous with respect to one goal may make good sense with respect to another. We saw this in chapter 5 in relation to social behaviour; the case of bet-hedging shows, unexpectedly, that it also arises in relation to choice under risk.

Finally, one technical point is worth clearing up. How does Grafen's point about expected relative fitness square with the observation in section 8.4.2, that ordering types by geometric mean fitness is not equivalent to ordering them by expected relative fitness? The answer is as follows. In the bet-hedging model, a strategy is optimal, by the geometric mean criterion, if and only if it maximizes expected relative fitness when fixed in the population. However, this equivalence says nothing about sub-optimal strategies; in particular, it does not say that if two sub-optimal strategies differ in geometric mean, the better one also does better by the expected relative fitness criterion.

This point is not merely a technicality; it also highlights the distinction between the two types of agential thinking. The issue about ordering in section 8.4.2 arose when we asked how nature 'chooses' between competing types in the face of risk; there the focus was on the process of selection and its putative analogy with an agent's choice. However, the bet-hedging issue arises in considering the evolved choices of an organism. In this context it makes sense to focus on the optimal strategy, since the point of agential thinking (type 1) is to provide a way of understanding adaptation, not natural selection.

8.7 Conclusion

Decision-theoretic ideas arise in two areas of biology: risk-sensitive foraging, which focuses on variability in resource-acquisition, and the theory of evolution in variable environments. The former concerns the (actual) choices of adapted organisms, the latter the (metaphorical) 'choices' made by natural selection. Here we have investigated whether both sorts of choices can be understood on the model of expected utility maximization, if utility is suitably defined in terms of biological fitness. In many scenarios, we find that they can, though the required definition is a somewhat subtle matter. In other scenarios, for example, involving arbitrary combinations of aggregate and idiosyncratic risk, we find that evolution and rationality part ways. The overall moral is that the connections between decision theory and evolutionary biology, though instructive, must be treated with care. Given the psychological pull of agential thinking, it is tempting to assume that the organism-as-rational-agent concept *must* be applicable, if organisms' choice behaviour has been optimized by natural selection. However, this has the status of theoretical hypothesis, not a priori truth.

Final Thoughts

This brings us to the end of the journey. The discussion has ranged quite widely, so it is worth stepping back to re-capitulate the main points and to extract some general morals.

Part I focused on a mode of thinking in evolutionary biology that we called 'agential'. This involves using notions such as interests, goals, and strategies in evolutionary analysis. Agential thinking has a number of manifestations. One is the use of intentional idioms ('wants, knows'), usually in an extended or metaphorical sense, to describe adaptive behaviour. Another is the analogical transfer of concepts from rational choice theory to evolutionary biology. There are two types of agential thinking, which need to be sharply distinguished. Type 1 treats an evolved entity, paradigmatically an individual organism, as akin to an agent with a goal towards which its phenotypic traits, including its behaviour, conduce. Type 2 treats 'mother nature', a personification of natural selection, as akin to a rational agent choosing between alternatives in accordance with a goal, such as maximal fitness. The former is a way of thinking about adaptation (the product), the latter about selection (the process).

The validity of the two types of agential thinking does not stand or fall together; this is a surprising finding, given that selection and adaptation are closely related concepts. Agential thinking (type 2) can be useful, but is potentially misleading. For it only really works well if we confine attention to the simplest sort of natural selection, in which the selective environment is constant and the complexities of inheritance are ignored. Agential thinking (type 1) is different. It is a valid expression of adaptationist assumptions, and its ultimate justification is empirical rather than theoretical. To the extent that an evolved organism is well-adapted to its environment, and thus equipped with phenotypic traits that enhance its survival and reproduction, it can be validly treated as agent-like as long as a certain empirical precondition is met, at least approximately. This is the unity-of-purpose condition: the organism's traits must have evolved because of their contribution to a *single* overall goal, so have complementary rather than antagonistic functions. To the extent that this is not so, it ceases to be possible to think of the organism as agent-like. This is a biological analogue of the psychological unity-of-purpose that is presupposed when we make intentional attributions to human agents.

The unity-of-purpose condition is not always met, as a certain amount of internal conflict does occur in many evolved organisms. The same is true, to a more significant extent, in most biological groups. This both points to a limitation of agential thinking (type 1) and highlights its rationale. It shows that the organism-as-agent concept is not idle metaphor, and nor is it merely a long-winded alternative to the standard talk of Darwinian function. For the functional idiom applies on a trait-by-trait basis, so is applicable whether or not organisms exhibit unity-of-purpose. Thus there is an objective pattern in nature that agential thinking (type 1) helps to capture, namely the organismic unity and integration that are presupposed by the practice of making agential and intentional attributions.

Part II examined the vexed issue of fitness-maximization in evolutionary biology, which is closely connected with agential thinking of both types. Intuitively both natural selection and adaptation have something to do with fitness being maximized, but whether this idea can be made precise, and if so how, has long been a matter of controversy. Two classical approaches to this issue, Wright's adaptive landscape and Fisher's fundamental theorem, focus on the effect of selection on the change and 'partial change', respectively, in mean population fitness. However, the former does not show that selection will always lead to hill-climbing, and the latter points only to an attenuated sense in which selection is an optimizing force. More recent approaches to the issue include Grafen's 'individual as maximizing agent' idea, Hamilton's idea of inclusive fitness maximization in social evolution, and the idea that frequency-dependent selection will lead to phenotypes that maximize individual fitness conditionally on being fixed in the population. However, each of these ideas has limitations and/or requires restrictive assumptions, so collectively they fall short of showing that fitness-maximization is a predicted outcome of evolution by natural selection.

This analysis has three main morals. First, there seems to be no general theoretical principle to the effect that natural selection will tend to produce adaptation, even in the absence of the 'constraints' and 'non-selective forces' that are usually invoked to explain non-adaptive outcomes. The grounds for being an adaptationist must ultimately be empirical, not theoretical. Second, this reinforces the independence of the two types of agential thinking. The infirmities of the doctrine of fitness-maximization show why it is misleading to assimilate the process of natural selection to a rational agent's deliberate choice between alternatives; but this leaves untouched the idea that a well-adapted organism can usefully be treated as an agent with a goal, to the extent that it exhibits the requisite unity-of-purpose. Third, in the context of social evolution theory, agential thinking (type 1) must be applied with caution, as the appropriate 'goal' depends on model assumptions. The most general version of Hamilton's rule, though always true, implies nothing about the quantity, if any, that organisms will appear to maximize in their social behaviour.

Part III focused on rationality, and tried to elucidate the murky relation between evolution and rational choice. There are two dimensions to this relation: the evolution of rationality itself, considered as an actual phenotypic attribute of some organisms; and the use of the organism-as-rational-agent heuristic as a tool for understanding adaptive behaviour. (The latter is a specific application of agential thinking (type 1) to behavioural traits.) These dimensions are separate but related; indicative of this is that certain formal models admit of a dual interpretation: either as explanations of how rationality actually evolved in those organisms that exhibit it, or of why rational idioms are well-suited to characterizing adaptive behaviour. The link between the two dimensions is explained by the hypothesis that adaptive behaviour is a form of proto-rationality.

One natural idea is that the capacity for rational thought and action is itself an adaptation that evolved by natural selection, as many authors believe. In broad terms this idea seems defensible. However, the more specific claim that natural selection would favour behaviour that satisfies standard definitions of rationality, for example, as codified in rational choice theory, is a different matter. This is probably true in many cases; but it is also conceivable that there may be an evolutionary explanation for why organisms, including humans, systematically *violate* rational choice norms; and arguments to this effect have been often been proposed. Such arguments pose a challenge to agential thinking (type 1), and suggest that no simple evolutionary naturalization of these rationality norms will work.

Given the similarity between the fitness-maximizing paradigm of evolutionary biology and the utility-maximizing paradigm of rational choice theory, it is tempting to suggest, as many authors have done, that adaptive and rational behaviour should coincide, if fitness is equated with utility. Often this is indeed the case; however, there exist numerous models that suggest that maximization of fitness and of utility can sometimes part ways, that is, that irrationality may be biologically advantageous. Though interesting, the parting-of-ways phenomenon is not fatal to the organism-as-rational-agent heuristic, as in some cases the irrationality is more apparent than real, and in other cases it can be eliminated by suitably re-framing the decision problem, or by suitable choice of utility function and/or fitness measure. The moral is that where evolved behaviour is concerned, agential thinking (type 1) must be employed with care; since when rationality is defined sufficiently precisely, it becomes an open theoretical question whether the adaptive and the rational will coincide, not a definitional truth. A concrete illustration of this moral comes from comparing the economic analysis of rational choice in the face of risk with the biological analysis of optimal behaviour in risky environments.

In the Introduction, we identified two overarching philosophical questions that would run throughout the book. The first was whether the prevalence of agential thinking in evolutionary biology is a reflection of objective biological facts, or of the human predilection to anthropomorphize. The answer is 'a bit of both'. On the one hand, there is a genuine point to treating an evolved organism as akin to an agent with a goal; organisms really do have attributes that make this mode of thinking apposite. On the other hand, agential thinking (type 2) is often misleading; and that of type 1 can also lead astray, for example, when its empirical preconditions are not met, or when inappropriate fitness measures are employed, or when the parting-of-ways possibility is ignored. So the attraction of agential thinking in biology is probably partly attributable to the peculiar purchase that thinking in terms of agents, goals, and strategies has on the human mind.

The second issue was how should we make sense of the joint facts that intentional and rational idioms are used to theorize about evolution, and that there is an evolutionary story to be told about how creatures with the cognitive capacities needed for intentionality and rationality evolved? The answer is 'by tracing the systematic connections between these facts'. This may sound glib, but I hope that the analysis of Part III has helped to flesh the answer out. If the story we have sketched about how the cognitive capacities in question evolved, it becomes intelligible why a set of concepts and descriptors whose original application is to the purposive behaviour of humans should be useful in adaptationist theorizing. Furthermore, there is a fairly intimate connection between the two dimensions of the evolution–rationality connection; indicative of this is that parting-of-ways arguments are typically relevant to both dimensions at once.

I am conscious that these answers are far from definitive, but that should be no surprise. For although the book is about science, and has tried to engage seriously with the scientific literature, its guiding questions are philosophical, and thus by their nature do not admit of final answers. I am also conscious that the way I have set the issues up is not the only possible way, and that my analysis could be improved on in numerous respects. But I hope that there is an element of truth in what I have written, and that the journey has been enjoyable for the reader.

References

Abbot, P. et al. (2011). Inclusive fitness theory and eusociality. *Nature*, 471:E1–E2.

Abrams, P. (2001). Adaptationism, optimality models, and tests of adaptive scenarios. In Orzack, S. H. and Sober, E., editors, *Adaptationism and Optimality*, pages 273–302. Cambridge University Press, Cambridge.

Agranov, M. and Ortoleva, P. (2017). Stochastic choice and preference for randomization. *Journal of Political Economy*, 125:40–68.

Ainslee, G. (1992). *Picoeconomics*. Cambridge University Press, Cambridge.

Alger, W. and Weibull, J. W. (2012). A generalization of Hamilton's rule—love others how much? *Journal of Theoretical Biology*, 299:42–54.

Allais, M. (1953). Le comportement de l'homme rationnel devant le risque: critique des postulats et axiomes de l'école américaine. *Econometrica*, 21:503–46.

Allais, M. and Hagen, O., editors (1979). *Expected Utility Hypothesis and the Allais Paradox*. Reidel, Dordrecht.

Allee, W. C. (1931). *Animal Aggregations. A Study in General Sociology*. Chicago University Press, Chicago, IL.

Allen, C. and Bekoff, M. (1999). *Species of Mind*. MIT Press, Cambridge, MA.

Allen, B. and Nowak, M. A. (2015). Games among relatives revisited. *Journal of Theoretical Biology*, 194:391–407.

Allen, B., Nowak, M. A., and Wilson, E. O. (2013). Limitations of inclusive fitness. *Proceedings of the National Academy of Sciences USA*, 110:21035–9.

Amundson, R. (2005). *The Changing Role of the Embryo in Evolutionary Thought*. Cambridge University Press, Cambridge.

Arnold, S. J., Pfrender, M. E., and Jones, A. G. (2001). The adaptive landscape as a conceptual bridge between micro- and macroevolution. *Genetica*, 1:9–32.

Autzen, B. (2017). Musing on means: fitness, expectation and the principles of natural selection. *British Journal for the Philosophy of Science*, axx031, https://doi.org/10.1093/bjps/axx031.

Barrett, J. L. (2004). *Why Would Anyone Believe in God?* AltaMira Press, Walnut Creek, CA.

Barrett, L. (2011). *Beyond the Brain*. Princeton University Press, Princeton, NJ.

Bateson, M. (2002). Recent advances in our understanding of risk-sensitive foraging preferences. *Proceedings of the Nutrition Society*, 61:509–16.

Bateson, P. and Laland, K. N. (2013). Tinbergen's four questions: an appreciation and an update. *Trends in Ecology and Evolution*, 28:712–18.

Bergstrom, T. C. (1995). On the evolution of altruistic ethical rules for siblings. *American Economic Review*, 85:58–81.

Bergstrom, T. C. (2014). On the evolution of hoarding, risk-taking and wealth distribution in nonhuman and human populations. *Proceedings of the National Academy of Sciences USA*, 111:10860–7.

Berninghaus, S., Güth, W., and Kliemt, H. (2012). Pull, push or both? Indirect evolution in economics and beyond. In Okasha, S. and Binmore, K., editors, *Evolution and Rationality*, pages 110–32. Cambridge University Press, Cambridge.

Bernoulli, D. ([1738] 1954). Exposition of a new theory on the measurement of risk. *Econometrica*, 22:23–36.

Binmore, K. (1987). Modeling rational players part 1. *Economics and Philosophy*, 3:179–214.

Birch, J. (2014). Hamilton's rule and its discontents. *British Journal for the Philosophy of Science*, 65:381–411.

Birch, J. (2016). Natural selection and the maximization of fitness. *Biological Reviews*, 91:712–27.

Birch, J. (2017). *The Philosophy of Social Evolution*. Oxford University Press, Oxford.

Birch, J. and Okasha, S. (2015). Kin selection and its critics. *Bioscience*, 65:22–32.

Bolton, G. E. and Ockenfels, A. (2000). ERC: a theory of equity, reciprocity, and competition. *The American Economic Review*, 90:166–93.

Boorse, C. (1975). On the distinction between disease and illness. *Philosophy and Public Affairs*, 5:49–68.

Bourke, A. F. G. (2011). *Principles of Social Evolution*. Oxford University Press, Oxford.

Bratman, M. (1987). *Intention, Plans, and Practical Reason*. Harvard University Press, Cambridge, MA.

Brennan, T. J. and Lo, A. W. (2011). The origin of behaviour. *Quarterly Journal of Finance*, 1:55–108.

Broome, J. (1991). *Weighing Goods*. Blackwell, Oxford.

Brown, P. M. (1976). Discussion: Conditionalization and expected utility. *Philosophy of Science*, 43:415–19.

Buchak, L. (2013). *Risk and Rationality*. Oxford University Press, Oxford.

Burt, A. and Trivers, R. (2006). *Genes in Conflict*. Harvard University Press, Cambridge, MA.

Buss, L. W. (1988). *The Evolution of Individuality*. Princeton University Press, Princeton, NJ.

Byrne, A. and Whiten, A. (1988). *Machiavellian Intelligence*. Oxford University Press, Oxford.

Calcott, B. and Sterelny, K. (2011). *The Major Transitions in Evolution Revisited*. MIT Press, Cambridge, MA.

Camerer, C. F. (2003). *Behavioral Game Theory*. Princeton University Press, Princeton, NJ.

Cant, M. A. and Johnstone, R. A. (2009). How threats influence the evolutionary resolution of within-group conflict. *American Naturalist*, 173:759–71.

Carruthers, P. (2006). *The Architecture of the Mind*. Oxford University Press, Oxford.

Charlesworth, B. (2008). *Evolution in Age-Structured Populations, 2nd edition*. Cambridge University Press, Cambridge.

Chater, N. (2012). Building blocks of human decision making. In Hammerstein, P. and Stevens, J. R., editors, *Evolution and the Mechanisms of Decision Making*, pages 53–68. Cambridge, MA, MIT Press.

Childs, D. Z., Metcalf, C. J., and Rees, M. (2010). Evolutionary bet-hedging in the real world: empirical evidence and challenges revealed by plants. *Proceedings of the Royal Society B*, 277:3055–64.

Clarke, E. (2014). Origins of evolutionary transitions. *Journal of Biosciences*, 39:303–17.

Clayton, N., Emery, E., and Dickinson, A. (2006). The rationality of animal memory. In Hurley, S. and Nudds, M., editors, *Rational Animals*, pages 197–216. Oxford University Press, Oxford.

Cooper, W. S. (2003). *The Evolution of Reason*. Cambridge University Press, Cambridge.

Cooper, W. S. and Kaplan, R. H. (1982). Adaptive 'coin-flipping': a decision-theoretical examination of natural selection for a random individual variation. *Journal of Theoretical Biology*, 94:135–51.

Coyne, J. A., Barton, N. H., and Turelli, M. (1997). A critique of Sewall Wright's shifting balance theory of evolution. *Evolution*, 51:643–71.

Curry, P. (2001). Decision making under uncertainty and the evolution of interdependent preferences. *Journal of Economic Theory*, 98:357–69.

Dall, S. R. X., Giraldeau, L., Olsson, O., McNamara, J. M., and Stephens, D. W. (2005). Information and its use by animals in evolutionary ecology. *Trends in Ecology and Evolution*, 20:187–93.

Damuth, J. and Heisler, I. L. (1988). Alternative formulations of multilevel selection. *Biology and Philosophy*, 3:407–30.

Danielson, P. A. (1998). *Modeling Rationality, Morality and Evolution*. Oxford University Press, Oxford.

Darwin, C. (1859). *On the Origin of Species by Means of Natural Selection*. John Murray, London.

Darwin, C. (1871). *The Descent of Man, and Selection in Relation to Sex*. John Murray, London.

Davidson, D. (2001a). *Essays on Actions and Events*. Oxford University Press, Oxford.

Davidson, D. (2001b). *Inquiries into Truth and Interpretation*. Oxford University Press, Oxford.

Dawkins, R. (1976). *The Selfish Gene*. Oxford University Press, Oxford.

Dawkins, R. (1981). In defence of selfish genes. *Philosophy*, 56:556–73.

Dawkins, R. (1982). *The Extended Phenotype*. Oxford University Press, Oxford.

Day, T. and Taylor, P. D. (1998). Unifying genetic and game theoretic models of kin selection for continuous traits. *Journal of Theoretical Biology*, 194:391–407.

de Sousa, R. (2007). *Why Think? Evolution and the Rational Mind*. Oxford University Press, Oxford.

Debove, S., Baumard, N., and Andre, J. (2016). Models of the evolution of fairness in the ultimatum game: a review and classification. *Evolution and Human Behaviour*, 37:245–54.

Dekel, E., Ely, J. C., and Yilankaya, O. (2007). Evolution of preferences. *The Review of Economic Studies*, 74:685–704.

Dennett, D. C. (1969). *Content and Consciousness*. Routledge, London.

Dennett, D. C. (1983). Intentional systems in cognitive ethology. *Behavioural and Brain Sciences*, 6:343–55.

Dennett, D. C. (1987). *The Intentional Stance*. MIT Press, Cambridge, MA.

Dennett, D. C. (1990). The interpretation of texts, people, and other artifacts. *Philosophy and Phenomenological Research*, 50:177–94.

Dennett, D. C. (1995). *Darwin's Dangerous Idea*. Penguin, London.

Dethlefsen, L., McFall-Nagi, M., and Relman, D. A. (2007). An ecological and evolutionary perspective on human-microbe mutualism and disease. *Nature*, 449:811–18.

Dieckmann, U. and Ferriere, R. (2004). Adaptive dynamics and evolving biodiversity. In Ferriere, R., Dieckmann, U., and Couvet, D., editors, *Evolutionary Conservation Biology*, pages 188–224. Cambridge University Press, Cambridge.

Diekmann, O. (2004). A beginner's guide to adaptive dynamics. In Rudnicki, R., editor, *Mathematical Modelling of Population Dynamics*, pages 47–86. Polish Academy of Sciences, Warsaw.

Dietrich, F. and List, C. (2016). Mentalism versus behaviourism in economics: a philosophy-of-science perspective. *Economics and Philosophy*, 32:249–81.

Doebeli, M. (2011). *Adaptive Diversification*. Princeton University Press, Princeton, NJ.

Drayson, Z. (2014). The personal/subpersonal distinction. *Philosophy Compass*, 9:338–46.

Dretske, F. (1998). *Explaining Behaviour*. MIT Press, Cambridge MA.

Durkheim, E. ([1895] 1982). *The Rules of Sociological Method*. Free Press, New York.

Easley, D. and Kleinberg, J. (2010). *Networks, Crowds and Markets*. Cambridge University Press, Cambridge.

Edwards, A. W. F. (1994). The fundamental theorem of natural selection. *Biological Reviews*, 69:443–74.

Edwards, A. W. F. (2002). The fundamental theorem of natural selection. *Theoretical Population Biology*, 61:335–7.

Edwards, A. W. F. (2014). R. A. Fisher's gene-centred view of evolution and the fundamental theorem of natural selection. *Biological Reviews*, 89:135–47.

Edwards, S. C. and Pratt, S. C. (2009). Rationality in collective decision-making by ant colonies. *Proceedings of the Royal Society B*, 276:3655–61.

Ellsberg, D. (1961). Risk, ambiguity, and the Savage axioms. *Quarterly Journal of Economics*, 75:643–69.

Elster, J. (1982). The case for methodological individualism. *Theory and Society*, 11:453–82.

Elster, J. (1985). *Making Sense of Marx*. Cambridge University Press, Cambridge.

Emerson, A. E. (1939). Social coordination and the superorganism. *American Midland Naturalist*, 21:182–209.

Eshel, I. (1983). Evolutionary and continuous stability. *Journal of Theoretical Biology*, 103: 99–111.

Ewens, W. J. (1989). An interpretation and proof of the fundamental theorem of natural selection. *Theoretical Population Biology*, 36:167–89.

Ewens, W. J. (2010). What is the gene trying to do? *British Journal for the Philosophy of Science*, 62:155–76.

Ewens, W. J. and Lessard, S. (2015). On the interpretation and relevance of the fundamental theorem of natural selection. *Theoretical Population Biology*, 104:59–67.

Falconer, D. S. (1985). A note on Fisher's 'average effect' and 'average excess'. *Genetical Research*, 46:337–47.

Falconer, D. S. and McKay, T. F. C. (1996). *Introduction to Quantitative Genetics, 4th edition*. Pearson, Cambridge.

Fawcett, T. W., McNamara, J. M., and Houston, A. I. (2012). When is it adaptive to be patient? A general framework for evaluating delayed rewards. *Behavioural Processes* 89, 128–36.

Fehr, E. and Schmidt, K. M. (1999). A theory of fairness, competition, and cooperation. *The Quarterly Journal of Economics*, 114:817–68.

Fishburn, P. C. (1991). Nontransitive preferences in decision theory. *Journal of Risk and Uncertainty*, 4:113–34.

Fishburn, P. C. and Rubinstein, A. (1982). Time preference. *International Economic Review*, 23:677–94.

Fisher, R. A. (1930). *The Genetical Theory of Natural Selection*. Clarendon Press, Oxford.

Fisher, R. A. (1941). Average excess and average effect of a gene substitution. *Annals of Eugenics*, 11:53–63.

Fodor, J. and Piattelli-Palmarini, M. (2010). *What Darwin got Wrong*. Profile Books, London.

Frank, S. A. (1998). *Foundations of Social Evolution*. Princeton University Press, Princeton, NJ.

Frank, S. A. (2003). Repression of competition and the evolution of cooperation. *Evolution*, 57:693–705.

Frank, S. A. (2013). Natural selection VII. History and interpretation of kin selection theory. *Journal of Evolutionary Biology*, 26:1151–84.

Frank, S. A. and Slatkin, M. (1990). Evolution in variable environments. *American Naturalist*, 136(2):244–60.

Frank, S. A. and Slatkin, M. (1992). Fisher's fundamental theorem of natural selection. *Trends in Ecology and Evolution*, 7(3):92–95.

Frankish, K. (2009). Systems and levels: dual systems theories and the personal-subpersonal distinction. In Evans, J. and Frankish, K., editors, *In Two Minds*, pages 89–107. Oxford University Press, Oxford.

Frederick, S., Loewenstein, G., and O'Donoghue, T. (2002). Time discounting and time preference: a critical review. *Journal of Economic Literature*, 40:351–401.

Futuyma, D. (2009). *Evolution*. Sinauer Associates, Sunderland, MA.

Gale, J., Binmore, K. G., and Samuleson, L. (1995). Learning to be imperfect: the ultimatum game. *Games and Economic Behaviour*, 8:56–90.

Gardner, A. and Grafen, A. (2009). Capturing the superorganism. *Journal of Evolutionary Biology*, 22:659–71.

Gardner, A. and Welch, J. (2011). A formal theory of the selfish gene. *Journal of Evolutionary Biology*, 24:1801–13.

Gardner, A., West, S.A., and Barton, N. (2007). The relation between multilocus population genetics and social evolution theory. *American Naturalist* 169:207–26.

Gardner, A., West, S. A., and Wild, G. (2011). The genetical theory of kin selection. *Journal of Evolutionary Biology*, 24:1020–43.

Gelman, A. and Hill, J. (2007). *Data Analysis using Regression and Multilevel/Hierarchical Models*. Cambridge University Press, Cambridge.

Geritz, S. A. H., Kisdi, E., Meszena, G., and Metz, J. A. J. (1998). Evolutionarily singular strategies and the adaptive growth and branching of the evolutionary tree. *Evolutionary Ecology*, 12:32–57.

Gershenson, S. (1928). A new sex-ratio abnormality in *Drosophila obscura*. *Genetics*, 13: 488–507.

Gigerenzer, G. and Selten, R. (2001). *Bounded Rationality*. MIT Press, Cambridge, MA.

Gigerenzer, G. and Todd, P. (1999). *Simple Heuristics that Make us Smart*. Oxford University Press, Oxford.

Gigerenzer, G., Todd, P., and Pachur, T., editors (2011). *Heuristics: the Foundations of Adaptive Behaviour*. Oxford University Press, Oxford.

Gillespie, J. H. (1974). Natural selection for within-generation variance in offspring number. *Genetics*, 76:601–6.

Gillespie, J. H. (1977). Natural selection for variances in offspring number: a new evolutionary principle. *American Naturalist*, 111:1010–14.

Gintis, H. (2009). *The Bounds of Reason*. Princeton University Press, Princeton, NJ.

Giraldeau, L. (1997). The ecology of information use. In Krebs, J. R. and Davies, N. B., editors, *Behavioural Ecology: an evolutionary approach*, pages 42–88. Blackwell, Oxford.

Godfrey-Smith, P. (1996). *Complexity and the Function of Mind in Nature*. Cambridge University Press, Cambridge.

Godfrey-Smith, P. (2009). *Darwinian Populations and Natural Selection*. Oxford University Press, Oxford.

Gomulkiewicz, R. (1998). Game theory, optimization and quantitative genetics. In Dugatkin, L. A. and Reeve, H. K., editors, *Game Theory and Animal Behaviour*, pages 283–303. Oxford University Press, Oxford.

Goodnight, C. J. (2012). Wright's shifting balance theory and the factors affecting the probability of peak shifts. In Svensson, E. and Calsbeek, R., editors, *The Adaptive Landscape in Evolutionary Biology*, pages 74–86. Oxford University Press, Oxford.

Gould, S. J. and Lewontin, R. C. (1979). The spandrels of San Marco and the Panglossian paradigm. *Proceedings of the Royal Society of London B*, 205:581–98.

Grafen, A. (1979). The hawk-dove game played between relatives. *Animal Behaviour*, 27:905–7.

Grafen, A. (1982). How not to measure inclusive fitness. *Nature*, 298:425–6.

Grafen, A. (1984). Natural selection, kin selection and group selection. In Krebs, J. R. and Davies, N. B., editors, *Behavioural Ecology, 2nd edition*, pages 62–84. Blackwell Scientific Publications, Oxford.

Grafen, A. (1999). Formal Darwinism, the individual-as-maximising-agent analogy, and bet-hedging. *Proceedings of the Royal Society B*, 266:799–803.

Grafen, A. (2002). A first formal link between the Price equation and an optimization program. *Journal of Theoretical Biology*, 217:75–91.

Grafen, A. (2003). Fisher the evolutionary biologist. *The Statistician*, 52:319–29.

Grafen, A. (2006). Optimization of inclusive fitness. *Journal of Theoretical Biology*, 238:541–63.

Grafen, A. (2008). The simplest formal argument for fitness optimization. *Journal of Genetics*, 87:421–33.

Grafen, A. (2014a). The formal Darwinism project in outline. *Biology and Philosophy*, 29:155–74.

Grafen, A. (2014b). The formal Darwinism project in outline: response to commentaries. *Biology and Philosophy*, 29:281–92.

Green, R. F. (1980). Bayesian birds: a simple example of Oaten's stochastic model of optimal foraging. *Theoretical Population Biology*, 18:244–56.

Güth, W. and Kliemt, H. (1998). The indirect evolutionary approach: bridging the gap between rationality and adaptation. *Rationality and Society*, 10:377–99.

Güth, W. and Kocher, M. G. (2013). More than thirty years of ultimatum game experiments. *Jena Economic Research Papers*, 35:1–36.

Güth, W. and Peleg, B. (2001). When will payoff maximization survive? An indirect evolutionary analysis. *Journal of Evolutionary Economics*, 10:479–99.

Güth, W. and Yaari, M. (1992). Explaining reciprocal behaviour in simple strategic games. In Witt, U., editor, *Explaining Process and Change: Approaches to Evolutionary Economics*, pages 23–34. University of Michigan Press, Ann Arbor, MI.

Hagen, E. H., Chater, N., Gallistel, C. R., Houston, A., Kacelnik, A., Kalenscher, T., Nettle, D., Oppenheimer, D., and Stephens, D. W. (2012). Decision making: what can evolution do for us? In Hammerstein, P. and Stevens, J. R., editors, *Evolution and the Mechanisms of Decision Making*, pages 97–128. MIT Press, Cambridge, MA.

Haig, D. (2002). *Genomic Imprinting and Kinship*. Rutgers University Press, New Brunswick, NJ.

Haig, D. (2004). Genomic imprinting and kinship: how good is the evidence? *Annual Review of Genetics*, 31:553–85.

Haig, D. (2006). Intrapersonal conflict. In Jones, M. K. and Fabian, A. C., editors, *Conflict*, pages 8–22. Cambridge University Press, Cambridge.

Haig, D. (2008). Kinship asymmetries and the divided self. *Behavioural and Brain Sciences*, 31:271–2.

Haig, D. (2012). The strategic gene. *Biology and Philosophy*, 27:461–79.

Haig, D. and Grafen, A. (1991). Genetic scrambling as a defense against meiotic drive. *Journal of Theoretical Biology*, 153:531–58.

Haldane, J. B. S. (1932). *The Causes of Evolution*. Longmans, Green and Co., London.

Haldane, J. B. S. (1955). Population genetics. *New Biology*, 18:34–51.

Hamilton, W. D. (1964). The genetical evolution of social behaviour. *Journal of Theoretical Biology*, 7:1–52.

Hamilton, W. D. (1971). Selection of selfish and altruistic behaviour in some extreme models. In *Narrow Roads of Gene Land, volume 1*, pages 198–228. W. H. Freeman, New York.

Hamilton, W. D. (1975). Innate social aptitudes in man: an approach from evolutionary genetics. In Fox, R. L., editor, *Biosocial Anthropology*, pages 115–32. John Wiley, New York.

Hammerstein, P. (2012). Towards a Darwinian theory of decision making. In Okasha, S. and Binmore, K., editors, *Evolution and Rationality*, pages 7–22. Cambridge University Press, Cambridge.

Hammond, P. J. (2005). Utility invariance in non-cooperative games. In Schmidt, U. and Traub, S., editors, *Advances in Public Economics: Utility, Choice and Welfare*, pages 31–50. Springer, Amsterdam.

Harsanyi, J. C. (1953). Cardinal utility in welfare economics and in the theory of risk-taking. *Journal of Political Economy*, 61:434–5.

Harsanyi, J. C. (1955). Cardinal welfare, individualistic ethics, and interpersonal comparisons of utility. *Journal of Political Economy*, 63:309–21.

Hart, S. (2005). An interview with Robert Aumann. *Macroeconomic Dynamics*, 9:683–740.

Heath, J. (2015). Methodological individualism. In Zalta, E. N., editor, *The Stanford Encyclopedia of Philosophy*. Stanford. https://plato.stanford.edu/archives/fall2015/entries/methodological-individualism/.

Heisler, I. L. and Damuth, J. (1987). A method for analyzing selection in hierarchically structured populations. *American Naturalist*, 130:582–602.

Hempel, C. (1965). *Aspects of Scientific Explanation*. Free Press, New York.

Hintze, A., Olson, R. S., Adami, C., and Hertwig, R. (2015). Risk sensitivity as an evolutionary adaptation. *Scientific Reports*, 5:8242.

Hölldobler, B. and Wilson, E. O. (2008). *The Superorganism*. W. W. Norton, London.

Houston, A. I. (2012). Natural selection and rational decisions. In Okasha, S. and Binmore, K., editors, *Evolution and Rationality*, pages 50–66. Cambridge University Press, Cambridge.

Houston, A. I., McNamara, J. M., and Steer, M. D. (2007a). Do we expect natural selection to produce rational behaviour? *Philosophical Transactions of the Royal Society B*, 362:1531–43.

Houston, A. I., McNamara, J. M., and Steer, M. D. (2007b). Violations of transitivity under fitness maximization. *Biology Letters*, 3:365–7.

Huck, S. and Müller, W. (2012). Allais for all: revisiting the paradox in a large representative sample. *Journal of Risk and Uncertainty*, 44:261–93.

Jeffrey, R. C. (1990). *The Logic of Decision, 2nd edition*. University of Chicago Press, Chicago, IL.

Kacelnik, A. (2006). Meanings of rationality. In Hurley, S. and Nudds, M., editors, *Rational Animals?*, pages 87–106. Oxford University Press, Oxford.

Kagel, J., Battalio, R., and Green, L. (1995). *Economic Choice Theory: an Experimental Analysis of Animal Behaviour*. Cambridge University Press, Cambridge.

Kahneman, D. (2011). *Thinking Fast and Slow*. Macmillan, London.

Kahneman, D. and Tversky, A. (1979). Prospect theory: an analysis of decision under risk. *Econometrica*, 67:263–91.

Kalenscher, T. and Pennartz, C. M. A. (2011). Do intransitive choices reflect genuinely context-dependent preferences? In Delgado, M. R., Phelps, E. A., and Robbins, T. W., editors, *Decision Making, Affect, and Learning: Attention and Performance XXIII*, pages 235–56. Oxford, Oxford University Press.

Kalenscher, T. and van Wingerden, M. (2011). Why we should use animals to study economic decision making—a perspective. *Frontiers in Neuroscience*, 5:82.

Kaplan, J. (2008). The end of the adaptive landscape metaphor? *Biology and Philosophy*, 23:625–38.

Karlin, S. and Matessi, C. (1983). Kin selection and altruism. *Proceedings of the Royal Society B*, 219:327–53.

Kelley, J. L. (1956). A new interpretation of information rate. *Bell System Technical Journal*, 35:917–26.

Kennett, J. and Matthews, S. (2003). The unity and disunity of agency. *Philosophy, Psychiatry and Psychology*, 10:308–12.

Kirkpatrick, M. and Rousset, F. (2005). Wright meets AD: not all landscapes are adaptive. *Journal of Evolutionary Biology*, 18:1166–9.

Kornblith, H. (2002). *Knowledge and its Place in Nature*. Oxford University Press, Oxford.

Korsgaard, C. M. (1989). Personal identity and the unity of agency. *Philosophy and Public Affairs*, 18:103–31.

Kreps, D. M. (1988). *Notes on the Theory of Choice*. Westview Press, Boulder, CO.

Kreps, D. M. and Wilson, R. (1982). Sequential equilibria. *Econometrica*, 50:863–94.

Lande, R. (1979). Quantitative genetic analysis of multivariate evolution, applied to brain-body size allometry. *Evolution*, 33:402–16.

Lande, R. (2007). Expected relative fitness and the adaptive topography of fluctuating landscapes. *Evolution*, 61:1835–46.

Lee, J. J. and Chow, C. C. (2013). The causal meaning of Fisher's average effect. *Genetical Research*, 95:89–109.

Lehmann, L., Alger, I., and Weibull, J. W. (2015). Does evolution lead to maximizing behaviour? *Evolution*, 69:1858–73.

Lehmann, L., Keller, L., West, S., and Roze, D. (2007). Group selection and kin selection: two concepts but one process. *Proceedings of the National Academy of Sciences USA*, 104:6736–9.

Lehmann, L. and Rousset, F. (2014). Fitness, inclusive fitness, and optimization. *Biology and Philosophy*, 29:181–95.

Lewis, D. (1981). Causal decision theory. *Australasian Journal of Philosophy*, 59:5–30.

Lewontin, R. (2000). *It Ain't Necessarily So*. New York Review Books, New York.

Lewontin, R. and Cohen, D. (1969). On population growth in a randomly varying environment. *Proceedings of the National Academy of Sciences USA*, 62:1056–60.

List, C. and Pettit, P. (2011). *Group Agents*. Oxford University Press, Oxford.

Luce, R. D. and Raiffa, H. (1957). *Games and Decisions*. John Wiley, New York.

Machina, M. (1982). 'Expected utility' analysis without the independence axiom. *Econometrica*, 50:277–323.

MacIntyre, A. (1999). *Dependent Rational Animals: Why Human Beings Need the Virtues*. Open Court, La Salle, IL.

Mallpress, D. E. W., Fawcett, T. W., Houston, A. I., and McNamara, J. M. (2015). Risk attitudes in a changing environment. *Psychological Review*, 122:364–75.

Marshall, J. A. (2011). Group selection and kin selection: formally equivalent approaches. *Trends in Ecology and Evolution*, 26:325–32.

Marshall, J. A. (2015). *Social Evolution and Inclusive Fitness Theory*. Princeton University Press, Princeton, NJ.

Martens, J. (2017). Inclusive fitness and the maximizing agent analogy. *British Journal for the Philosophy of Science*, 68:875–905.

Mas-Colell, A., Whinston, M. D., and Green, J. R. (1995). *Microeconomic Theory*. Oxford University Press, Oxford.

Maynard Smith, J. (1977). Parental investment: a prospective analysis. *Animal Behaviour*, 25:1–9.

Maynard Smith, J. (1978). Optimization theory in evolution. *Annual Review of Ecology and Systematics*, 9:31–56.

Maynard Smith, J. (1982). *Evolution and the Theory of Games*. Cambridge University Press, Cambridge.

Maynard Smith, J. and Harper, D. (2003). *Animal Signals*. Oxford University Press, Oxford.

Maynard Smith, J. and Price, G. R. (1973). The logic of animal conflict. *Nature*, 246:15–18.

Maynard Smith, J. and Szathmáry, E. (1995). *The Major Transitions in Evolution*. Oxford University Press, Oxford.

Mayr, E. (1988). *Towards a New Philosophy of Biology*. Harvard University Press, Cambridge, MA.

Mayr, E. (2009). *What Makes Biology Unique?* Cambridge University Press, Cambridge.

McDermott, R., Fowler, J. H., and Smirnov, O. (2008). On the evolutionary origin of prospect theory preferences. *The Journal of Politics*, 70:335–50.

McDowell, J. (1994). *Mind and World*. Harvard University Press, Cambridge MA.

McFarland, D. (2016). *The Biological Bases of Economic Behaviour*. Palgrave Macmillan, Basingstoke.

McNamara, J. M. (1995). Implicit frequency-dependence and kin selection in fluctuating environments. *Evolutionary Ecology*, 9:185–203.

McNamara, J. M. and Dall, S. (2010). Information is a fitness-enhancing resource. *Oikos*, 119:231–6.

McNamara, J. M., Green, R. F., and Olsson, O. (2006). Bayes' theorem and its applications in animal behaviour. *Oikos*, 112:243–51.

McNamara, J. M. and Houston, A. I. (2002). Credible threats and promises. *Philosophical Transactions of the Royal Society of London B*, 357:1607–16.

McNamara, J. M., Trimmer, P. C., and Houston, A. I. (2014). Natural selection can favour 'irrational' behaviour. *Biology Letters*, 10:20130935.

Metz, J. A. J. (2012). Adaptive dynamics. In Hastings, A. and Gross, L. J., editors, *Encyclopedia of Theoretical Ecology*, pages 7–17. California University Press, Oakland, CA.

Metz, J. A. J., Mylius, S. D., and Diekmann, O. (2008). When does evolution optimize? *Evolutionary Ecology Research*, 10:629–54.

Michod, R. E. (1999). *Darwinian Dynamics*. Princeton University Press, Princeton, NJ.

Midgley, M. (1979). Gene-juggling. *Philosophy*, 54:439–58.

Millikan, R. (2000). Reading Mother Nature's mind. In Ross, D., Brook, A., and Thompson, D., editors, *Dennett's Philosophy: a Comprehensive Assessment*, pages 55–76. MIT Press, Cambridge, MA.

Mishra, S. (2014). Decision-making under risk: integrating perspectives from biology, economics and psychology. *Personality and Social Psychology Review*, 18, 280–307.

Mitchell, W. A. and Valone, T. J. (1990). The optimization research program: studying adaptations by their function. *The Quarterly Review of Biology*, 65:43–52.

Mongin, P. (2001). The impartial observer theorem of social ethics. *Economics and Philosophy*, 17:147–9.

Monod, J. (1973). *Chance and Necessity*. Vintage Books, New York.

Moran, P. A. P. (1964). On the non-existence of adaptive topographies. *Annals of Human Genetics*, 27:383–93.

Mylius, S. D. and Diekmann, O. (1995). When does evolution optimize? *Oikos*, 74:218–24.

Nissen, L. A. (1997). *Teleological Language in the Life Sciences*. Rowman and Littlefield, Lanham, MD.

Nowak, M. (2006). *Evolutionary Dynamics*. Harvard University Press, Cambridge, MA.

Nowak, M. A., Tarnita, C. E., and Wilson, E. O. (2010). The evolution of eusociality. *Nature*, 466:1057–62.

Nunney, L. (1985). Group selection, altruism, and structured-deme models. *The American Naturalist*, 126:212–30.

Odling-Smee, F. J., Laland, K. N., and Feldman, M. W. (2003). *Niche-construction: The Neglected Process in Evolution*. Princeton University Press, Princeton, NJ.

O'Donald, P. (2000). Review of A. W. F. Edwards, *Foundations of Mathematical Genetics*. *Heredity*, 84:620–1.

Okasha, S. (2002). Genetic relatedness and the evolution of altruism. *Philosophy of Science*, 69:138–49.

Okasha, S. (2005). Multilevel selection and the major transitions in evolution. *Philosophy of Science*, 72:1013–25.

Okasha, S. (2006). *Evolution and the Levels of Selection*. Oxford University Press, Oxford.

Okasha, S. (2008). Fisher's fundamental theorem of natural selection—a philosophical analysis. *British Journal for the Philosophy of Science*, 59:319–51.

Okasha, S. (2012). Social justice, genomic justice and the veil of ignorance. *Economics and Philosophy*, 28:43–71.

Okasha, S. (2015). The relation between kin and multilevel selection. *British Journal for the Philosophy of Science*, 67:435–70.

Okasha, S. (2016). On the interpretation of decision theory. *Economics and Philosophy*, 32:409–33.

Okasha, S. (forthcoming). A note on the definition of fitness in stochastic environments.

Okasha, S. and Martens, J. (2016a). Hamilton's rule, inclusive fitness maximization, and the goal of individual behaviour in symmetric two-player games. *Journal of Evolutionary Biology*, 29:473–82.

Okasha, S. and Martens, J. (2016b). The causal meaning of Hamilton's rule. *Royal Society Open Science*, 29:160037.

Okasha, S. and Paternotte, C. (2012). Group adaptation, formal Darwinism, and contextual analysis. *Journal of Evolutionary Biology*, 25:1127–39.

Okasha, S. and Paternotte, C. (2014). Adaptation, fitness, and the selection-optimality links. *Biology and Philosophy*, 29:225–32.

Pearl, J. (2000). *Causality*. Cambridge University Press, Cambridge.

Pearl, J. (2001). Direct and indirect effects. In *Proceedings of the Seventeenth Conference on Uncertainty in Artificial Intelligence*, pages 411–20. Morgan Kaufmann Publishers, San Francisco, CA.

Peña, J., Nöldeke, G., and Lehmann, L. (2015). Evolutionary dynamics of collective action in spatially structured populations. *Journal of Theoretical Biology*, 382:122–36.

Pettigrew, R. (2016). *Accuracy and the Laws of Credence*. Oxford University Press, Oxford.

Price, G. R. (1970). Selection and covariance. *Nature*, 227:520–1.

Price, G. R. (1972a). Extension of covariance selection mathematics. *Annals of Human Genetics*, 35:485–90.

Price, G. R. (1972b). Fisher's 'fundamental theorem' made clear. *Annals of Human Genetics*, 36:129–40.

Proctor, D., Williamson, R. A., de Waal, F. B. M., and Brosnan, S. F. (2013). Chimpanzees play the ultimatum game. *Proceedings of the National Academy of Sciences USA*, 110:2070–5.

Proulx, S. R. and Day, T. (2001). What can invasion analyses tell us about evolution under stochasticity? *Selection*, 2:1–16.

Provine, W. B. (2001). *The Origins of Theoretical Population Genetics*. University of Chicago Press, Chicago.

Queller, D. C. (1984). Kin selection and frequency dependence. *Biological Journal of the Linnean Society*, 23:133–43.

Queller, D. C. (1992). Quantitative genetics, inclusive fitness and group selection. *American Naturalist*, 139:540–58.

Queller, D. C. (1997). Cooperators since life began. *Quarterly Review of Biology*, 72:184–8.

Queller, D. C. (2011). Expanded social fitness and Hamilton's rule for kin, kith and kind. *Proceedings of the National Academy of Sciences USA*, 108:10792–9.

Queller, D. C. and Strassmann, J. E. (2009). Beyond society: the evolution of organismality. *Philosophical Transactions of the Royal Society of London B*, 364:3143–55.

Queller, D. C. and Strassmann, J. E. (2013). The veil of ignorance can favour biological cooperation. *Biology Letters*, 9:20130365.

Quiggin, J. (1982). A theory of anticipated utility. *Journal of Economic Behaviour and Organisation*, 3:323–43.

Quine, W. V. O. (1969). Epistemology naturalized. In *Ontological Relativity and Other Essays*, pages 69–90. Columbia University Press, New York.

Ramsey, F. P. (1931). Truth and probability. In *The Foundations of Mathematics and Other Logical Essays*, pages 156–98. Routledge and Kegan Paul, London.

Rand, D. G., Tarnita, C. E., Ohtsuki, H., and Nowak, M. A. (2013). Evolution of fairness in the one-shot anonymous ultimatum game. *Proceedings of the National Academy of Sciences USA*, 110:2581–6.

Ratnieks, F. L. W., Foster, K. R., and Wenseleers, T. (2006). Conflict resolution in insect societies. *Annual Review of Entomology*, 51:581–608.

Rawls, J. (1971). *A Theory of Justice*. Harvard University Press, Cambridge, MA.

Real, L. A. (1992). Information processing and the evolutionary ecology of cognitive architecture. *The American Naturalist*, 140:S108–S145.

Real, L. A. and Caraco, T. (1986). Risk and foraging in stochastic environments. *Annual Review of Ecology and Systematics*, 17:371–90.

Rescorla, M. (2013). Rationality as a constitutive ideal. In Lepore, E. and Ludwig, E., editors, *A Companion to Donald Davidson*, pages 472–88. Blackwell, Oxford.

Rice, S. (2008). A stochastic version of the Price equation reveals the interplay of deterministic and stochastic processes in evolution. *BMC Evolutionary Biology*, 8:262.

Rice, S. H. (2004). *Evolutionary Theory*. Sinauer, Sunderland, MA.

Ridley, M. (2000). *Mendel's Demon*. Weidenfeld and Nicolson, London.

Ridley, M. and Grafen, A. (1981). Are green beard genes outlaws? *Animal Behaviour*, 29:954–5.

Ristau, C. (1991). Aspects of the cognitive ethology of an injury-feigning bird, the piping plover. In Ristau, C., editor, *Cognitive Ethology*, pages 93–124. Lawrence Erlbaum Associates, Hillsdale, NJ.

Roberts, F. S. (1985). *Measurement Theory*. Cambridge University Press, Cambridge.

Robson, A. (1996). A biological basis for expected and non-expected utility. *Journal of Economic Theory*, 68:397–424.

Robson, A. (2002). Evolution and human nature. *Journal of Economic Perspectives*, 16:89–106.

Robson, A. and Samuelson, L. (2012). The evolutionary foundations of preferences. In Bisin, A. and Jackson, M., editors, *Handbook of Social Economics*, pages 221–310. Amsterdam, North-Holland.

Roff, D. A. (1992). *Evolution of Life Histories*. Chapman and Hall, New York.

Ross-Gillespie, A. and Kummerli, R. (2014). Collective decision-making in microbes. *Frontiers in Microbiology*, 5:54.

Rousset, F. (2015). Regression, least squares, and the general version of inclusive fitness. *Evolution*, 69:2963–70.

Rovane, C. (1998). *The Bounds of Agency*. Princeton University Press, Princeton.

Russell, S. J. and Norvig, P. (1995). *Artificial Intelligence*. Prentice Hall, New Jersey.

Salmon, W. (1989). *Four Decades of Scientific Explanation*. University of Minnesota Press, Minneapolis, MN.

Samuelson, L. (1997). *Evolutionary Games and Equilibrium Selection*. MIT Press, Cambridge, MA.

Samuelson, L. (1988). Evolutionary foundations of solution concepts for finite two-player normal form games. In Vardi, M., editor, *Proceedings of the Second Conference on Theoretical Aspects of Reasoning about Knowledge*, pages 211–26. Morgan Kaufmann, Los Altos, CA.

Samuelson, P. A. (1937). A note on measurement of utility. *Review of Economic Studies*, 4:155–61.

Samuelson, P. A. (1971). The 'fallacy' of maximizing the geometric mean in long sequences of investing or gambling. *Proceedings of the National Academy of Sciences USA*, 68:2493–6.

Savage, L. J. (1954). *The Foundations of Statistics*. Wiley, New York.

Schlosser, M. (2015). Agency. In Zalta, E. N., editor, *The Stanford Encyclopedia of Philosophy*. Stanford. https://plato.stanford.edu/archives/fall2015/entries/agency/.

Schumm, G. F. (1987). Transitivity, preferences and indifference. *Philosophical Studies*, 53:435–7.

Scott-Phillips, T. C., Dickins, T. E., and West, S. A. (2011). Evolutionary theory and the ultimate/proximate distinction in the human behavioural sciences. *Perspectives on Psychological Science*, 6:38–47.

Seeley, T. D. (1996). *The Wisdom of the Hive*. Harvard University Press, Cambridge, MA.

Seeley, T. D. (2010). *Honeybee Democracy*. Princeton University Press, Princeton.

Seger, J. and Brockmann, H. J. (1987). What is bet-hedging? *Oxford Surveys in Evolutionary Biology*, 5:182–211.

Selten, R. (1975). Reexamination of the perfectness concept of equilibrium in extensive games. *International Journal of Game Theory*, 4:25–55.

Sen, A. (1976). Welfare inequalities and Rawlsian axiomatics. *Theory and Decision*, 7:243–62.

Sen, A. (1977a). Non-linear social welfare functions: a reply to Professor Harsanyi. In Butts, R. and Hintikka, J., editors, *Foundational Problems in the Special Sciences*, pages 297–302. Reidel, Dordrecht.

Sen, A. (1977b). On weights and measures: informational constraints in social welfare analysis. *Econometrica*, 45:1539–72.

Sen, A. (2009). *The Idea of Justice*. Penguin, London.

Shackleton, K., Al Toufailia, H., Balfour, N., Nascimento, F., Alves, D., and Ratnieks, F. (2015). Appetite for self-destruction: suicidal biting as a nest defense strategy in trigona stingless bees. *Behavioural Ecology and Sociobiology*, 69:273–81.

Shafir, S. (1994). Intransitivity of preferences in honey-bees. *Animal Behaviour*, 48:55–67.

Shafir, S., Waite, T. A., and Smith, B. (2002). Context-dependent violations of rational choice in honey-bees (*Apis mellifera*) and gray jays (*Perisoreus canadensis*). *Behavioural Ecology and Sociobiology*, 51:180–7.

Simpson, G. G. (1953). *The Major Features of Evolution*. Columbia University Press, New York.

Skyrms, B. (1995). *Evolution of the Social Contract*. Cambridge University Press, Cambridge.

Snell-Rood, E. C. (2013). An overview of the evolutionary causes and consequences of behavioural plasticity. *Animal Behaviour*, 85:1004–11.

Sober, E. (1998). Three differences between evolution and deliberation. In Danielson, P., editor, *Modeling Rationality, Morality and Evolution*, pages 408–22. Oxford University Press, Oxford.

Sober, E. (2001). The two faces of fitness. In Singh, R. S., Krimbas, C. B., Paul, D. B., and Beatty, J., editors, *Thinking about Evolution: Historical, Philosophical and Political Perspectives*, pages 309–21. Cambridge University Press, Cambridge.

Sozou, P. D. (1998). On hyperbolic discounting and uncertain hazard rates. *Philosophical Transactions of the Royal Society of London B*, 265:2015–20.

Spurrett, D. (2016). Does intragenomic conflict predict intrapersonal conflict? *Biology and Philosophy*, 31:313–33.

Starmer, C. (2000). Developments in non-expected utility theory. *Journal of Economic Literature*, 38:332–82.

Starrfelt, J. and Kokko, H. (2012). Bet-hedging—a triple trade-off between means, variances and correlations. *Biological Reviews of the Cambridge Philosophical Society*, 87:742–55.

Stephens, D. W. (1981). The logic of risk-sensitive foraging preferences. *Animal Behaviour*, 29:628–9.

Stephens, D. W. (2002). Discrimination, discounting and impulsivity: a role for an informational constraint. *Philosophical Transactions of the Royal Society of London B*, 357:1527–37.

Stephens, D. W. and Krebs, J. R. (1987). *Foraging Theory*. Princeton University Press, Princeton, NJ.

Sterelny, K. (2003). *Thought in a Hostile World*. Blackwell, Oxford.

Sterelny, K. (2012). From fitness to utility. In Okasha, S. and Binmore, K., editors, *Evolution and Rationality*, pages 246–73. Cambridge University Press, Cambridge.

Sterelny, K. and Kitcher, P. (1988). The return of the gene. *Journal of Philosophy*, 85:339–60.

Stevens, J. R. and Stephens, D. W. (2010). The adaptive nature of impulsivity. In Madden, G. J. and Bickel, W. K., editors, *Impulsivity: the Behavioural and Neurological Science of Discounting*, pages 361–87. APA Press, Washington, DC.

Strassmann, J. E. and Queller, D. C. (2007). Insect societies as divided organisms. *Proceedings of the National Academy of Sciences USA*, 104:8619–26.

Strotz, R. H. (1956). Myopia and inconsistency in dynamic utility maximization. *The Review of Economic Studies*, 23:165–80.

Svensson, E. and Calsbeek, R., editors (2012). *The Adaptive Landscape in Evolutionary Biology*. Oxford University Press, Oxford.

Taylor, C. and Nowak, M. A. (2007). Transforming the dilemma. *Evolution*, 61:2281–92.

Taylor, P. D., Wild, G., and Gardner, A. (2007). Direct fitness or inclusive fitness: how shall we model kin selection? *Journal of Evolutionary Biology*, 20:201–9.

Tinbergen, N. (1963). On the aims and methods of ethology. *Zeitschrift fur Tierpsychologie*, 20:410–33.

Todd, P. M. and Gigerenzer, G. (2012). *Ecological Rationality*. Oxford University Press, Oxford.

Trewavas, A. (2015). *Plant Behaviour and Intelligence*. Oxford University Press, Oxford.

Trivers, R. (2009). Genetic conflict within the individual. In *Sonderdruck der Berliner-Brandenburgische Akademie der Wissenschschaften 14*, pages 149–99. Akademie Verlag, Berlin.

Trivers, R. L. and Hare, H. (1976). Haplodiploidy and the evolution of the social insects. *Science*, 191:249–63.

Tversky, A. (1975). A critique of expected utility theory: descriptive and normative considerations. *Erkenntnis*, 9:163–73.

Urban, M. C., Burger, R., and Bolnick, D. I. (2013). Asymmetric selection and the evolution of extraordinary defences. *Nature Communications*, 4:2085.

Valone, T. J. (2006). Are animals capable of Bayesian updating? *Oikos*, 112:252–9.

van Baalen, M. and Rand, D. A. (1998). The unit of selection in viscous populations and the evolution of altruism. *Journal of Theoretical Biology*, 193:631–48.

van Veelen, M. (2009). Group selection, kin selection, altruism and cooperation: when inclusive fitness is right and when it can be wrong. *Journal of Theoretical Biology*, 259:589–600.

von Hayek, F. (1942). Scientism and the study of society I. *Economica*, 9:267–91.

von Neumann, J. and Morgenstern, O. (1944). *Theory of Games and Economic Behaviour*. Princeton University Press, Princeton.

Waddington, C. H. (1957). *The Strategy of the Genes*. Ruskin House, London.

Wade, M. J. (2012). Wright's adaptive landscape: testing the predictions of his shifting balance theory. In Svensson, E. and Calsbeek, R., editors, *The Adaptive Landscape in Evolutionary Biology*, pages 58–71. Oxford University Press, Oxford.

Wade, M. J. and Kalisz, S. (1990). The causes of natural selection. *Evolution*, 44:1947–55.

Wakker, P. P. (1994). Separating marginal utility and probabilistic risk aversion. *Theory and Decision*, 36:1–44.

Wakker, P. P. (2010). *Prospect Theory*. Cambridge University Press, Cambridge.

Walsh, D. M. (2015). *Organisms, Agency, and Evolution*. Cambridge University Press, Cambridge.

Weber, M. (1922). *Economy and Society*. University of California Press, Berkeley, CA.

Weibull, J. (1995). *Evolutionary Game Theory*. MIT Press, Cambridge, MA.

Weirich, P. (1986). Expected utility and risk. *British Journal for the Philosophy of Science*, 37:419–42.

West, S. A. and Gardner, A. (2013). Adaptation and inclusive fitness. *Current Biology*, 23:R577–R584.

West, S. A., Griffin, A. S., and Gardner, A. (2007). Social semantics: altruism, cooperation, mutualism, strong reciprocity and group selection. *Journal of Evolutionary Biology*, 20:415–32.

Weymark, J. (1991). A reconsideration of the Harsanyi-Sen debate on utilitarianism. In Elster, J. and Roemer, J. E., editors, *Interpersonal Comparisons of Well-being*, pages 225–320. Cambridge University Press, Cambridge.

Whitman, D. W. and Agrawal, A. A. (2009). What is phenotypic plasticity and why is it important? In Whitman, D. W. and Ananthankrishnan, T. N., editors, *Phenotypic Plasticity of Insects: Mechanisms and Consequences*, pages 1–63. Enfield, NH., Science Publisher.

Wiley, R. H. (2013). Signal detection, noise, and the evolution of communication. In Brumm, H., editor, *Animal Communication and Noise*, pages 7–30. Springer, Amsterdam.

Williams, G. C. (1966). *Adaptation and Natural Selection*. Princeton University Press, Princeton, NJ.

Wilson, D. S. and Sober, E. (1989). Reviving the superorganism. *Journal of Theoretical Biology*, 136:337–56.

Wilson, R. A. (2005). *Genes and the Agents of Life*. Cambridge University Press, Cambridge.

Wright, L. (1976). *Teleological Explanations*. University of California Press, Berkeley, CA.

Wright, S. (1922). Coefficients of inbreeding and relationship. *The American Naturalist*, 56:330–8.

Wright, S. (1932). The role of mutation, inbreeding, crossbreeding and selection in evolution. *Proceedings of the Sixth Annual Congress of Genetics*, 1:356–66.

Wright, S. (1949). Adaptation and selection. In Mayr, E., editor, *Genetics, Paleontology, and Evolution*, pages 365–89. Princeton University Press, Princeton, NJ.

Index

The manufacturer's authorised representative in the EU for product
safety is Oxford University Press España S.A. of El Parque Empresarial
San Fernando de Henares, Avenida de Castilla, 2 - 28830 Madrid
(www.oup.es/en or product.safety@oup.com). OUP España S.A. also acts
as importer into Spain of products made by the manufacturer.
Printed and bound by CPI Group (UK) Ltd, Croydon, CR0 4YY
05/10/2026
02227967-0009